ALABAMA GOLD

ALABAMA GOLD

A HISTORY OF THE SOUTH'S LAST MOTHER LODE

PEGGY JACKSON WALLS

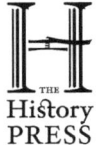

Published by The History Press
Charleston, SC
www.historypress.net

Copyright © 2016 by Peggy Jackson Walls
All rights reserved

First published 2016

ISBN 9781531698775

Library of Congress Control Number: 2016931935

Notice: The information in this book is true and complete to the best of our knowledge. It is offered without guarantee on the part of the author or The History Press. The author and The History Press disclaim all liability in connection with the use of this book.

All rights reserved. No part of this book may be reproduced or transmitted in any form whatsoever without prior written permission from the publisher except in the case of brief quotations embodied in critical articles and reviews.

Dedicated to the Hog Mountain gold miners whose stories made this book possible and especially to my father, Kermit Roosevelt Jackson (1909–1951).

Contents

Preface	9
Acknowledgements	13
Introduction	15

PART I: THE OLD SOUTHWEST: AMERICA'S FIRST GOLD RUSHES

1. "Alabama Fever"	25
2. Tallapoosa County's Gold Mining Districts: Devil's Backbone, Eagle Creek, Goldville and Hog Mountain	35
3. "It's Good to Be Shifty in a New Country"	47
4. Cotton Boom and Bust, Lost Confederate Gold, New Interest in Gold Mining	57
5. Tallapoosa County: "Gold Country"	71
6. Hillabee Gold Mining Company (1890–1916)	81

PART II: SURVIVING THE DEPRESSION: "GRINDING STONE INTO BREAD"

7. The Hog Mountain Mining and Milling Company: 1933–1937	91
8. Life in a Gold Mining Community	123
9. Notable People and Events	135
10. From the Mine to the Mill: J.P. Mooney	157

Appendix: An Incomplete List of Hog Mountain Gold Miners	161
Notes	163
Bibliography	167
Index	171
About the Author	175

Preface

In 1982, I began what would become a lifetime of researching, interviewing and writing about the Hog Mountain gold mine in northeast Tallapoosa County. Located less than three miles from my childhood home and rising four hundred feet from the surrounding landscape, the mountain was a daily, visible presence as familiar to my family as the sun rising in the east and setting in the west. I first learned the gold mining history of the mountain from conversations with family members and neighbors who worked for the Tallapoosa Mining and Milling Company in the 1930s. In school, I learned the geography and history of Tallapoosa County, where in the early 1800s, major battles took place between Native Americans and federal troops preparing the way for an influx of white settlers in the 1830s. But the Alabama history book contained no stories of America's first gold miners, who traveled down the Appalachian Mountains into the foothills of Georgia and crossed into Alabama.

Gold within the Alabama gold belt counties lay beneath the soil of the old Creek Nation, similar to deposits in Georgia, mainly found under Cherokee land. The geography and the chronology of events suggest the presence of gold might have contributed to the agitation leading to the Indian Removal Act in 1832. Important events in Alabama's history were the Native American culture, the Indian Wars of 1813-14 and the shifting of power from the Native Americans to the federal government after the defeat of the Creeks at the Battle of Horseshoe Bend on March 27, 1814. Included in the major events were the sale of the Native Americans' land to white settlers

Preface

and the evacuation of Indians from their homeland to the West. During the march, which became known as the Trail of Tears, thousands of Native Americans died from starvation, disease and exposure to the elements. The Five Civilized Tribes, listed as the Creek, Choctaw, Chickasaw, Seminoles and the Cherokee, assimilated into the white man's way of living—residing in cabins and growing vegetables—but they were nevertheless banished from their ancestral lands.

As ownership of the land in east-central Alabama was shifting in 1832, gold mining was taking place. But miners failed to keep personal records of their activities, or those who might have had access to the records did not deem them important enough to be preserved. Journals and papers of early travelers in Alabama, detailed conversations and letters of the military leaders, Indian agents and Native Americans survived from this period. Although rich in data regarding the Indian wars, they contained almost nothing about gold mining. Any records from this period were embedded with other information. From these scattered sources, *Alabama Gold* constructs and shares the story of antebellum gold mining in Alabama.

The introduction summarizes the stories of America's first gold discoveries in the southern Piedmont region: Cabarrus County, North Carolina; Lumpkin County, Georgia; and Chilton, Cleburne and Tallapoosa Counties in Alabama. The narrative reports of Alabama's first state geologist, Michael Tuomey, who served from 1848 to 1858, provide estimations of gold mines' potential in the Alabama gold belt and descriptions of mining strategies employed by the miners and mining companies. The surveys are largely technical but offer observations on the frenzied, disorganized manner in which antebellum mining was conducted. They describe hillsides pocked with holes that miners dug in haste and abandoned when they failed to find ore quickly. Recorded also are brief conversations with prospectors and local people, knowledgeable about mining activities in their areas. Without the careful notes of geologists, only a negligible amount of information about gold mining in Alabama would exist.

With the encouragement of Judge C.J. Coley, Governor John Patterson, Dr. J. Wayne Flynt, Dr. Leah Atkins, Dr. Bert Hitchcock, Dr. Jerry Brown and Dr. Patrick Morrow, in the 1980s, I continued to research, write and promote the history of gold mining at Hog Mountain.

A great deal of my research was primary, having grown up near Hog Mountain, I was acquainted with gold miners who worked during the Depression. They were of my parents' generation, and some were extended family. I found their stories fascinating and historically significant. I began

Preface

with the goal of publishing an article about the Hog Mountain Mining and Milling Company operation. This goal was accomplished in the publication of "Gold Mining at Hog Mountain in the 1930s" in the *Alabama Review* of July 1984. The twenty-five or so, interviews I conducted at this time also provided material on which I developed my master's thesis, "Folklore and Folk Life in Southern Prose Fiction," in which I linked Old Southwest humorist Johnson Jones Hooper's writing to life in the 1840s, when gold mining was at its peak in Tallapoosa County. In writing about stereotypical characters in southern fiction, I learned a few famous characters, such as Mark Twain's King and William Faulkner's Abner Snopes, might have had their genesis in the stereotypical con man Simon Suggs. Suggs was modeled after the frontier lawyer Bird Young, who was also an early settler of Youngsville, now Alexander City. The literary connections are interesting because the characters and stories that came out of the Old Southwest territory reemerged later in famous works of southern fiction. The story of antebellum gold mining is set in the Old Southwest, where danger was imminent due to hostilities between the white settlers and the Indians who viewed them as "intruders" on their land. During the early gold mining days, east-central Alabama was in transition from being the seat of the once powerful Creek Nation to embracing the early settlers and "gold diggers." *Alabama Gold* tells the story of antebellum gold mining and the pre–Civil War culture. The pre–World War I and Depression-era operations were motivated by economic circumstances and the American dream that anyone who persistently works hard will be rewarded with success. *Alabama Gold* is a history of all the miners who "busted and shoveled rock" to, first, gain a foothold in the young state of Alabama and, last, to hold on to it through the Depression.

Acknowledgements

I want to begin with an acknowledgement of academic mentors and friends Dr. J. Wayne Flynt and Dr. Leah Atkins, who encouraged me to continue researching and writing long after I completed my graduate work at Auburn University and promoted my research in their published work. I also would like to recognize Dr. Bert Hitchcock, Dr. Jerry Brown, Madison Jones, Nancy Anderson and Dr. Patrick Morrow for their support and encouragement.

In Tallapoosa County, Judge C.J. Coley made sure I had copies of all his historical papers, including his last copy of the *History of Tallapoosa County*, and introduced me to Dr. Ed Bridges and the valuable resources of the Alabama Department of Archives and History in Montgomery. Thank you, Governor John Patterson and Tina, for your wonderful support and encouragement

To all who opened their family albums, records and documents and made them accessible for use in *Alabama Gold*, thank you. Among them, John F. Farrow shared records and pictures of the Farrow Gold Mining Company's operations in southern Tallapoosa County. I regret that I cannot provide the names of all the miners and their families, but the story of *Alabama Gold* is their story as well. Thanks to Ted and Shirley Spears, who invited me to share the gold mining story at the B.B. Comer Memorial Library in Sylacauga, and to the Adelia Russell Library in Alexander City. The Tallapoosee County Historical Association and Museum in Dadeville provided access to valuable documents and pictures related to Tallapoosa County gold mining. Thank you to Coy Powell for guiding me and my youngest son, Eddie, through the

Acknowledgements

old caves and sites in Goldville in the 1980s, when I conducted most of the interviews in this book.

Thanks to my children, Bill and Melissa, and my grandchildren, Tyler, Tatum, Emma and Sophie. Your interest and encouragement kept me committed to completing the story of gold mining at Hog Mountain for the descendants of the miners and for a broad audience of historians and gold mining enthusiasts. Thanks to the members of the Ballard, Nelson and Baker families and so many others for sharing a piece of the mining history.

Special thanks go to Betty and Paul Wellborn for their permission to use pictures of Hog Mountain property and for sharing documents. Acknowledgement of current Hog Mountain pictures go to Audra Williams's Photography. A note of appreciation is extended to Phillip Padgett, Alton Padgett, Douglas Champion, Larkin Radney and Ben Russell, who contributed to the final collection of information. Appreciation for permission to use pictures is extended to the Pine Mountain Museum in Villa Rica, Georgia; the Tallapoosee County Historical Museum in Dadeville, Alabama; and the Alabama Department of Archives and History in Montgomery, Alabama.

Introduction

Discovery of gold in the southern Piedmont states preceded the California gold rush by several decades. In the 1820s, '30s and '40s, prospectors, land investors, politicians, farmers, itinerant preachers, conmen and outright scoundrels made their way into America's new frontiers, where the gold fields in Virginia, the Carolinas, Tennessee, Georgia and Alabama were teeming with mining activity. Southern gold was formed during volcanic and tectonic activity millions of years ago at the same time as the Appalachian Mountains. Deep within the Appalachian range, an intermittent line of gold veins, known as the Wedowee Schist, extended as far south as the Appalachian foothills of east-central counties in Alabama, such as Talladega, Randolph and Tallapoosa.

Myths of Indian chiefs with hordes of hidden treasures attracted the first gold seekers to America's southeastern region. With plans to claim the land and treasures for the Spanish empire, Alonso Álvarez de Pineda and his men arrived at present-day Mobile Bay in 1519. Pánfilo de Narváez and his party reached the Gulf Coast in 1527. Hernando de Soto, with seven hundred men and servants, two hundred horses, a drove of hogs and a pack of bloodhounds, came ashore on the west coast of present-day Florida in 1539. Each of these explorers, Alonso Álvarez de Pineda, Pánfilo de Narváez and Hernando de Soto, and their fellow conquistadors failed in their mission to build a wealthy Spanish empire in the New World. De Soto and his men traveled throughout the southeast in Florida, Georgia, the Carolinas, Tennessee and other Appalachian regions.

Introduction

In their first encounters with De Soto and his men, Native Americans welcomed the Europeans with food and gifts. The Spaniards responded by looting and destroying their villages and taking natives to serve as slaves and guides. In their three-year expedition, the conquistadors destroyed the Native Americans' centuries-old way of life. Tens of thousands of natives died after being exposed to European diseases against which they had no immunity. Entire tribes deserted villages to escape death. The Spaniards did not fare well after depleting the natives' food supply. After exhausting their own food supplies, they became so hungry they ate their horses. Sickness, starvation and battles with the Native Americans killed half of the conquistadors. Hernando de Soto's expedition ended in the near mutiny of the remaining soldiers. On March 21, 1542, Hernando de Soto died of fever beside the Mississippi River, which he is credited with discovering. Following de Soto's death, the Spaniards abandoned the failed expedition in the interior of the American wilderness and returned home. For all the death and destruction Hernando de Soto and his men brought to the Native Americans, they failed to discover gold or to establish a Spanish colony. Thus ended the Spaniards' search for gold in the southern Piedmont and the Native Americans' first encounters with Europeans.[1]

The Spaniards diminished the Native Americans' ability to defend themselves and their land against the French and the English who came into the southern Appalachian region in the eighteenth and nineteenth centuries. In the early nineteenth century, Europeans came in great numbers to the southeastern territory and succeeded where Hernando de Soto failed. They broke the power of the southeastern Indian tribes in the Indian Wars of 1813–14, claimed their land for the federal government and turned the Native Americans' ancestral hunting lands into farming land and mining fields. They discovered gold in some of the same regions Hernando de Soto and his men passed through in the southern Piedmont states of Virginia, the Carolinas, Tennessee, Georgia and Alabama.

The story of pioneer gold mining in the southern Piedmont begins in Virginia with one of the nation's founding fathers: signer of the Declaration of Independence, third president, planter, architect and author Thomas Jefferson.

Introduction

Virginia Gold

In 1782, Thomas Jefferson documented the discovery of gold in Virginia in his *Notes on the State of Virginia* (1785): "I knew a single instance of gold found in this state. It was interspersed in small specks through a lump of ore of about four pounds weight, which yielded seventeen pennyweights of gold." The ore was found on the north side of Rappahannock River.[2]

Palmer C. Sweet, in *Gold in Virginia*, reported the first lode deposit was discovered at the Whitehall mine in western Spotsylvania County. The first incorporated gold-mining company was the Virginia Mining Company of New York, operating in Orange County. Gold production from 1804 to 1828, consisting of an estimated 121 troy ounces, had an approximate value of $2,500. In the years from 1840 to 1849, production averaged nearly 3,000 ounces annually. Other southern Piedmont states followed the downward trend in gold production after gold was discovered in the western frontier in 1848. This trend continued through the Civil War years.[3]

North Carolina Gold

The first major gold discovery in the southern Piedmont took place in Cabarrus County, North Carolina. In 1799, twelve-year-old Conrad Reed, son of John and Sarah, noticed a yellow rock protruding from Little Meadow Creek where he was fishing. He retrieved the rock and showed it to his father, who was unable to identify its composition. The Reed family used the rock as a doorstop for three years. Finally, John Reed took the rock to a Fayetteville jeweler to be examined. Not realizing its value, he offered to sell the rock to the jeweler for $3.50, who gladly paid the sum. A year later, John Reed recovered about $1,000 from the sale. Reed continued to farm but also started a mining enterprise. He took three men—Frederick Kiser, James Love and Martin Phifer Jr.—into a partnership. One of James Love's slaves, Peter, discovered a twenty-eight-pound nugget, valued at $6,600. By 1824, sporadic mining activity on the Reed farm yielded about $100,000 worth of gold. Learning of the discoveries of gold on the Reed farm, other farmers began to dig in their fields, hills and streams, finding gold nuggets of different sizes and values. Gold mining in North Carolina attracted thousands of gold seekers.

More than a decade before the word "gold" was associated with California, North Carolina was known as "the golden state."[4] By 1832, North Carolina

Introduction

mines employed more than twenty-five thousand people. In the same year, the first US gold dollar coin was minted in North Carolina at the private home of Christopher Bechtler. His business minted $770,000 in gold coins from August 1836 to May 1838.[5]

South Carolina Gold

Gold was discovered in 1802 in Greenville in the Carolina Slate Belt, extending in a northeast–southeast direction in the same geologic formation as found in North Carolina. In 1829, in Camden, Lancaster County, the largest discovery of gold in South Carolina occurred at the Haile mine. The profit-sharing mine paid $1.50 to $3.00 a day to one to two hundred miners. Gold was shipped to the Philadelphia mint. When gold was discovered in California, a large exodus of miners in the Carolinas reduced mining activity. The downward trend of gold production continued until the Civil War, when President Lincoln ordered General Sherman to destroy the facilities.[6]

Georgia Gold

Gold strikes were reported in Georgia as early as 1823 near Milledgeville and in 1826 at Villa Rica in Carroll County. But "little attention was paid to these discoveries at the time."[7]

In 1828, a hunter, Benjamin Parks, discovered an unusual-looking rock near LickLog in the present Lumpkin County. A gold rush followed with the discovery of gold veins richer than those found in North Carolina. During the Georgia gold rush, LickLog's name was changed to Dahlonega, the Cherokee word for "yellow money." Unfortunately for the Cherokees, their land was the site of precious ores, which might have contributed to Jackson's Indian Removal Act to clear the way for gold miners and cotton farmers. Indian Removal from Cherokee Land coincides with the discovery of gold, though Europeans had long desired to possess the land of the Cherokees due to the developing cotton agriculture. In 1838–39, the Cherokees were removed from their ancestral lands to Indian Territory in Oklahoma.[8]

Introduction

Alabama Gold

Anticipating discoveries of gold in Alabama similar to those experienced in other Piedmont states, prospectors crossed into Cleburne County along the Villa Rica plateau. Since gold was discovered in the neighboring state of Georgia in 1828, gold was believed to have been discovered in Alabama in 1830.[9] Although the amount of gold recovered was less than that found in Georgia, even a small amount of money infused into the economy of farming-mining communities stimulated growth and excited miners with the possibility of becoming rich. The story of gold mining in Alabama is intertwined with the broader settlement history, Indian wars and Indian removal to the West. Decades before a western cowboy strapped on his "peacemaker," frontiersmen and women were "fighting Indians" and staking claim to land in the Old Southwest territory. They lived in rowdy, crowded camps in a hostile environment, where they might be killed or cheated out of their claim or any gold they recovered. From this scene, tall tales found their way into the writing of Old Southwest humorists, such as Johnson Jones Hooper.

Few records exist to tell about the exciting, peak years of antebellum gold mining in Alabama. But the geological surveys and reports of Alabama' first state geologist, Michael Tuomey, provide a glimpse of gold mining activity during the early to mid-1800s. Thus Alabama's earliest gold mining story was left to the statistics and data of Alabama's first geologists and to the imagination of Old Southwest writers, such as Hooper, Tallapoosa County's first census taker in 1840. Journals and papers of early travelers in Alabama and detailed conversations and letters of the military leaders, Indian agents and even Native Americans survived from this period, but they share almost no information about the gold mining activity in Alabama that included at least two gold rushes.

Thousands of settlers came down the Appalachian Mountains as far south as Alabama to claim newly evacuated Creek land and establish farms. Others came to find the "yellow stuff" rumored to be hidden in Alabama hills and streams. Many believed the Indians knew about rich gold deposits and kept this knowledge concealed from the Europeans. But if they knew where gold was concealed, they neglected its use in artifacts, preferring instead to use copper, stone and shells. The malleability and durability of copper were desirable qualities for molding arrowheads, breastplates, jewelry and ornaments. Large displays of Indian artifacts at the Alabama Department of Archives and History in Montgomery, Horseshoe Bend National Military Park in Tallapoosa County and other museums illustrate

Introduction

the region's rich Native American heritage and their use of natural resources to craft practical tools, artistic vases and ornamental jewelry. Neither is the gold mining history reflected in Alabama museums; it exists, instead, in the folk history of the mining districts, the records of geologists and the imagination of modern mining enthusiasts.

Geologist George I. Adams contended that gold mining was "an important factor in the early history of Alabama."[10] The prospect of finding gold attracted rugged individuals with the determination to wrest wealth and good fortune from the land. Prospectors worked day and night with lanterns digging in fields, streams and clay hills of the piney backwoods, searching for enough "yellow stuff" to make them wealthy. Farmers were sometimes rewarded with small nuggets or flakes they discovered while fishing, farming or hunting. The amount of gold retrieved from Alabama hills and streams in the early 1800s will never be known since much of the gold never made it to a mint but was used locally to purchase supplies, food, livestock and equipment. Alabama miners took an undetermined amount of southern gold to grubstake their claims in the western gold fields as they, with many other southern Piedmont miners, joined the California gold rush in 1849. The story of gold mining at Hog Mountain begins in the distant past when the Old Southwest was America's frontier, the scene of Indian wars, gold rushes, violation of treaties and Indian removal. The federal land offices were busy establishments with legal and fraudulent claims being filed on lands ceded by Indian treaties; some were legitimate purchases, but other land purchases were the result of tricking a Native American out of land or outright stealing of property.

The story of Alabama's most productive gold mine, Hog Mountain, begins with the settlers, prospectors and land speculators who found their way down the Appalachian Mountains into foothills of east-central Alabama. They were Old Southwest frontier families who settled the region, the miners who sought riches in gold and their descendants who, from 1839 to the present, have kept alive an interest in gold mining at Hog Mountain. With highlights from state geological surveys and mining, engineering and history journals and snippets of gold mining events from old newspapers, such as the *Niles Register* and the *Vidette*, *Alabama Gold* shares the story of early gold mining in Alabama in the 1800s. Old documents include a letter written from James Dowd Phillips from the California gold field to his wife, Sarah Ann, in Tallapoosa County describing the 1849 gold rush mining in California. Phillips owned the Blue Hill gold mine in the Devil's Backbone district in Tallapoosa County.

Introduction

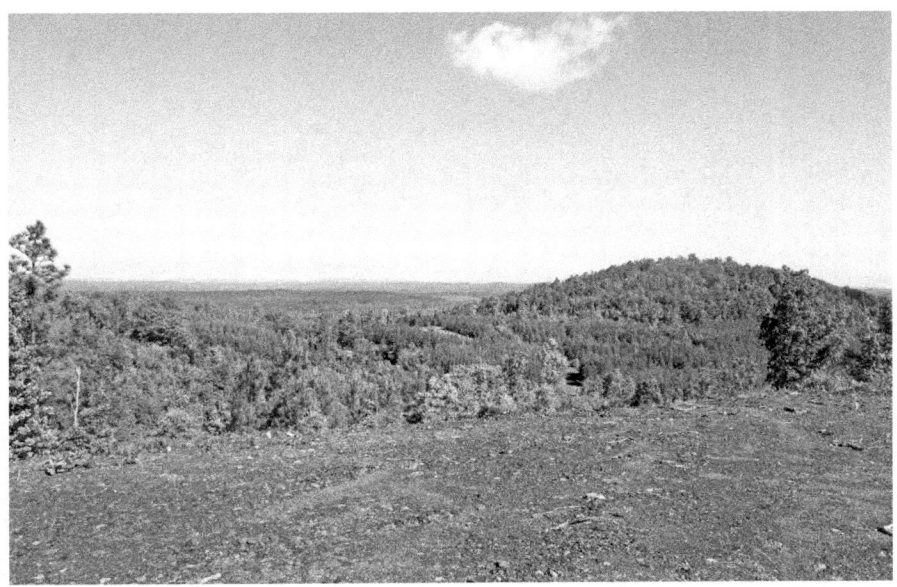

This panoramic view of Hog Mountain and surrounding landscape in 2014 shows the location of three successful former gold mining operations. *Courtesy of Peggy Jackson Walls.*

In the words of Depression-era miners, *Alabama Gold* shares the unique story of the Hog Mountain gold mining in the 1930s. The story did not end with the closing of the Hog Mountain Mining and Milling operation (1933–37). The old caretaker, known only as Mr. Neil, built a house cropping out over Hillabee Creek, where he spent his days washing out gold from the Hog Mountain tailings pile. The tradition of gold mining was continued by locals who panned in the streams of Hillabee, Enitachopco and Broken Arrow and gold mining enthusiasts who joined them. The stories of the miners are expanded and supported with statistical information from geological surveys written by George I. Adams, T.H. Aldrich Jr., William Phillips, E.A. Smith, William Brewer and plant superintendent N.O. Johnson.

Part I

THE OLD SOUTHWEST: AMERICA'S FIRST GOLD RUSHES

The significance of the southern excitement both as pilot experiment and education institution should not be minimized, not to mention the impetus it might well have given the land-speculation, canal-building, wildcat-banking boom of the mid-1830s. It is arguable that this lamentably speculative bubble may well have been founded upon the comfortable supposition that all would come right when the paper promises of the banks and of the states were redeemed by Appalachian gold. Yet the industry itself was of southern origins, and it introduced Americans to the idea of rushing for gold and to the techniques with which a hard-working man...might with a little luck gain comfortable wealth.
—historian Otis Young

1
"Alabama Fever"

Southern Piedmont gold was formed during volcanic and tectonic activity hundreds of millions of years ago at the same time as the Appalachian Mountains. The igneous and metamorphic rocks, known as the Wedowee Schist, followed a general southwest–northeast trend in an irregular line deep within the mountains. They extended as far south as the foothills of east-central Alabama into the present-day counties of Talladega, Randolph and Tallapoosa, located in the Alabama gold belt. Other counties that are partially or wholly in the 3,500-square-mile triangle of land are present-day Chambers, Chilton, Clay, Coosa, Cleburne and Elmore. But at the turn of the nineteenth century, the land known now as Alabama was a part of the Mississippi Territory and belonged to the Creek Nation, which held sovereign power over their ancestral lands. But change was underway that would bring thousands of settlers and miners to the Old Southwest frontier. Early historians found no evidence that the Spanish explorers or the Native Indians discovered gold and consider it likely that "intruders" on the Creek land were the first to discover gold in Alabama. State geologist George I. Adams suggested a connection between the dates of the discovery of gold in 1830 and agitation for removal of Indians in 1832. No definitive document exists to establish precisely the year gold was discovered in Alabama, but geologists and researchers have suggested following the discovery of gold in Georgia in 1829, prospectors would be anxious to follow the veins into Alabama in their search for more gold deposits. Geography and chronology support their view.[11]

Alabama Gold

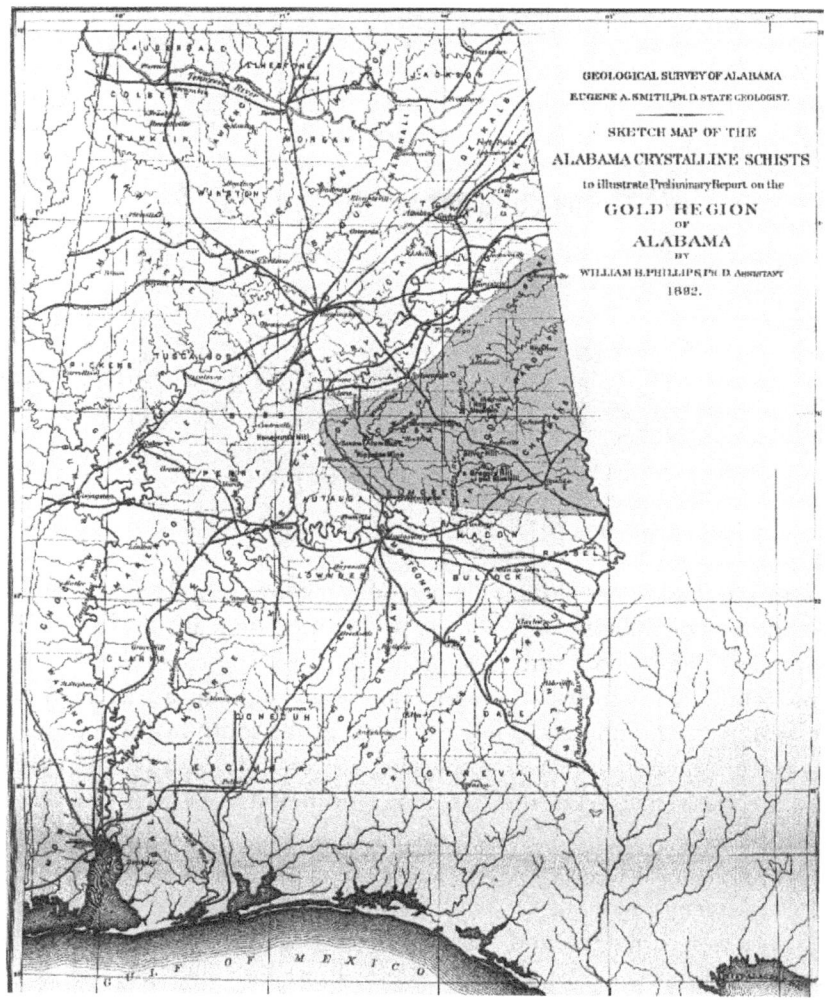

A sketch of a map of the gold region in Alabama in 1892, prepared for William B. Phillips's report on Alabama Crystalline Schists. *Courtesy of the Alabama Department of Archives and History.*

The Old Federal Road: A Thoroughfare Through Creek Land

In 1806, during Thomas Jefferson's presidency, a postal path was opened for riders to travel from middle Georgia to lower Alabama. Extending deep into Muscogee land, the primitive road became a thoroughfare for travelers, military troops and equipment. In 1811, the path was broadened into what became known as the Old Federal Road for the conveyance of commerce, military weapons and equipment. Forts, inns and stations were built near the road to accommodate travelers and troops. Native Americans watched with increasing anxiety as large numbers of military troops and weapons passed on the road. Tensions soon erupted into the 1813–14 Indian wars.

THE MASSACRE AT FORT MIMS

Many historians regarded the wars as a continuation of the War of 1812 and a civil war between two factions of Creeks. Indians who assimilated into the white man's lifestyle were targets for the warriors known as Red Sticks, whose name stemmed from their practice of painting their rifles red to symbolize war. Skirmishes between settlers and Indians increased in violence as the Creek attempted to defend their land and food supply from the "intruders," and settlers were equally determined to take possession of the land. The July 27, 1813 Battle of the Burnt Corn Creek is considered by many to be the first battle of the Creek Wars of 1813–14. Notified by a spy who observed Peter McQueen and a few Red Stick warriors as they purchased gunpowder in Mobile, Captain Dixon Baily McQueen and his militia lay in wait at Burnt Corn Creek and attacked the Red Sticks as they ate their noon meal. Few lives were lost, but the entire stock of gunpowder and other supplies were confiscated. In retaliation, on August 30, 1813, Peter McQueen and "Red Eagle" Weatherford led approximately 700 Red Stick warriors in an attack on Fort Mims, killing 250 men, women and children and taking 150 prisoners. The Fort Mims Massacre was the first major battle in the Indian Wars of 1813–14.[12]

MASSACRE AT HILLABEE TOWN

Benjamin Hawkins (1754–1816) was the principal Indian agent for the southeastern region. He was particularly successful in working with the Hillabee tribes in present-day northeast Tallapoosa County and Clay County, where he visited with Robert Grierson, a Scottish trader and cattleman. The name *Hillabee*, also spelled *Hillabi* and *Hilibis*, represented a

ALABAMA GOLD

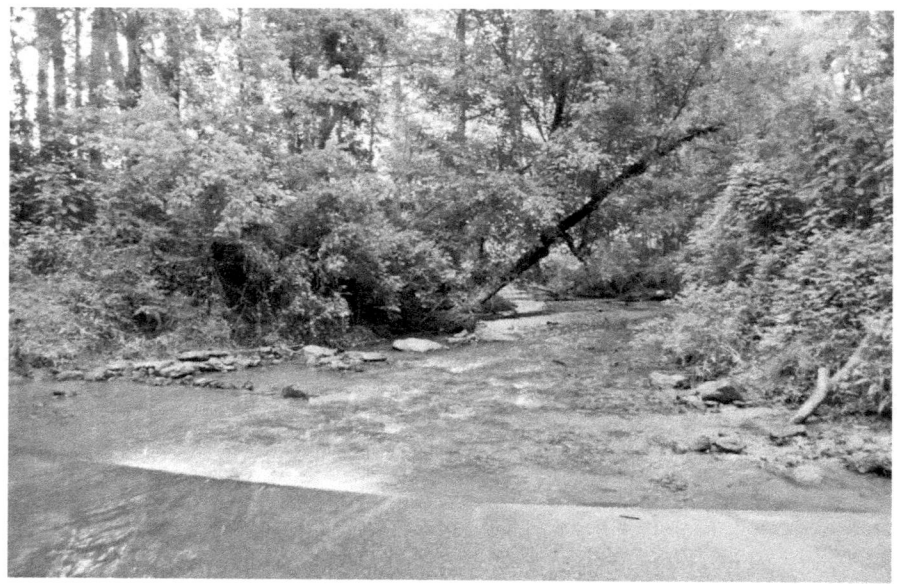

Broken Arrow Creek, named for the Hillabee Tribes who lived along the Hillabee Creek in northeast Tallapoosa County. *Courtesy of Peggy Jackson Walls.*

small group of villages on Hillabee Creek and Enitachopco (Anatitchapko). Grierson and his family lived on a large farm near the main Hillabee Town and natives who had assimilated into the white culture, living in cabins and growing cotton, wheat and vegetables. His wife, Sinnuggee, was a member of the Spanalge clan of the Hilibis. As the husband of a native, Grierson was protected and moved safely among the Indians conducting business. His farming and cattle raising were successful, and on his plantation, he had one of the first cotton gins made by Eli Whitney. Indians lived near Hillabee Creek, from which they derived their tribe name. Grierson sought and received amnesty for the peaceful tribe from General Andrew Jackson, but Major General John Cocke of the Tennessee Militia, unaware of the amnesty granted to the friendly tribe, dispatched General James White to destroy the Hillabee villages on November 18, 1813. Several hundred Indians were taken prisoners, and sixty-four, including women, were killed. As a result of the attack, the Hillabee Indians joined the Red Stick faction in fighting Andrew Jackson at Horseshoe Bend.

The Indian Wars of 1813–14 ended at the Battle of Horseshoe Bend on March 27, 1814. General Andrew Jackson, with his 3,300 soldiers and allied Cherokees, attacked 1,100 Red Stick warriors gathered at Tohopeka. By the end of the day, the Tallapoosa River ran red with the blood of 800 Native Americans killed in battle. Jackson's troops sustained only a few casualties, although a few men died of their injuries later. The battle broke the power of the Creek Nation and secured the vast territory and its rich resources for the United States. When Mississippi was officially recognized as a state in 1817, Alabama became a territory. William Bibb served as governor of the territory, becoming the first governor of Alabama when it was recognized by the US government as a state in 1819. Alabama was carved from the twenty-two million acres of the Creek land ceded to the federal government in the Treaty of Fort Jackson. Alexander "Sandy" Grierson, son of Robert Grierson, was one of the signers.

The year after Andrew Jackson became president (1829–37), he created the Indian Removal Law.[13] Creeks signed a treaty in March 1832 that opened a large portion of their Alabama land to white settlement but guaranteed them protected ownership of the remaining portion, which was divided among the leading families. The government did not protect them from speculators, however, who quickly cheated many out of their lands. By 1835, the destitute Creeks began stealing livestock and crops from white settlers. Some eventually committed arson and murder in retaliation for their brutal treatment. In 1836, the secretary of war ordered the removal of the Creeks as a military necessity. By 1837, approximately fifteen thousand Creeks had migrated west. They had never signed a removal treaty.[14]

Alabama Fever

The expression "Alabama Fever" was in use prior to Alabama's becoming a state due to Europeans' desire to possess a piece of the rich soil that was so well suited to grow cotton. After the Indian Removal Act was enforced, settlers rushed to stake their claims; they squatted on land not yet vacated by the Native Americans. Settlers came in oxen-drawn wagons bulging with household items and farm animals moving alongside them. Farmers and miners, eager to claim the land to grow cotton and claim any precious minerals they might unearth, came on horseback and on mules with farming and mining equipment. The first settlers were determined to grab a piece of the Old Southwest frontier, even if it meant driving the Native Americans from their ancestral homeland. Many immigrants settled in east-central Alabama, where crystalline rocks were abundant and where, with primitive

Alabama Gold

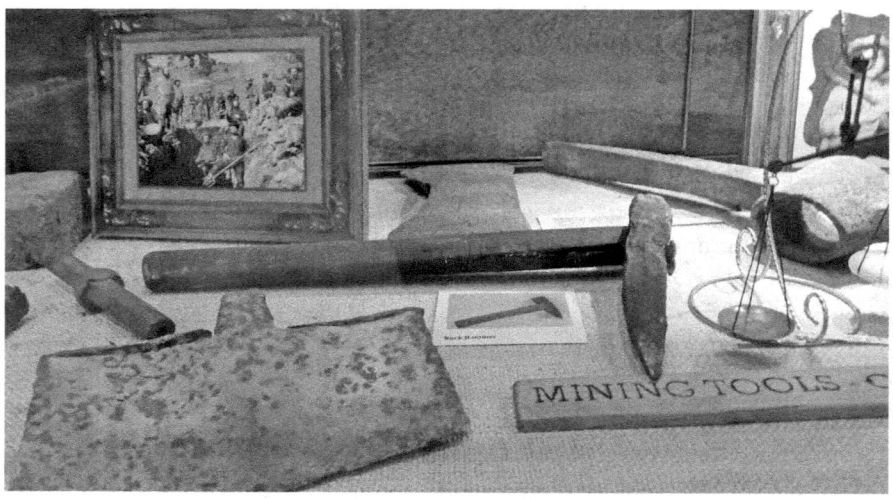

This picture of mining tools and miners is used with permission of Pine Mountain Gold Museum, located in Villa Rica, Georgia. *Courtesy of Peggy Jackson Walls.*

tools, miners removed placer gold from the ground. Using alluvial mining methods, they retrieved gold flakes and small nuggets from natural traps in the streams and eddies of free-flowing streams.

Early Gold Mining in Alabama: 1830-49

In 1829 and 1830, miners moved westward from Georgia into Alabama, mining the Villa Rica placer that extended into Cleburne County. Unfortunately, Alabama gold was located on land belonging to the Creek Nation in Alabama, just as Georgia gold was located on the Cherokees' ancestral land. After the removal of the Indians in Georgia, the Cherokee land was distributed by lotteries in which any non-Cherokee household could participate; in Alabama, deeds for Creek land were sold at federal land offices by government agents. The first authenticated discovery of gold in Alabama was made in 1830 by the nephew and namesake of Governor William Wyatt Bibb and his father-in-law, Todd Robinson. Their discovery of gold in old Autauga County (now Chilton County) was reported in the *Niles Herald*:[15]

> *Wyatt and Robinson collected their first gold on a tributary of Chestnut Creek, part of an eighty-acre tract they had purchased in the fall of 1830*

THE OLD SOUTHWEST

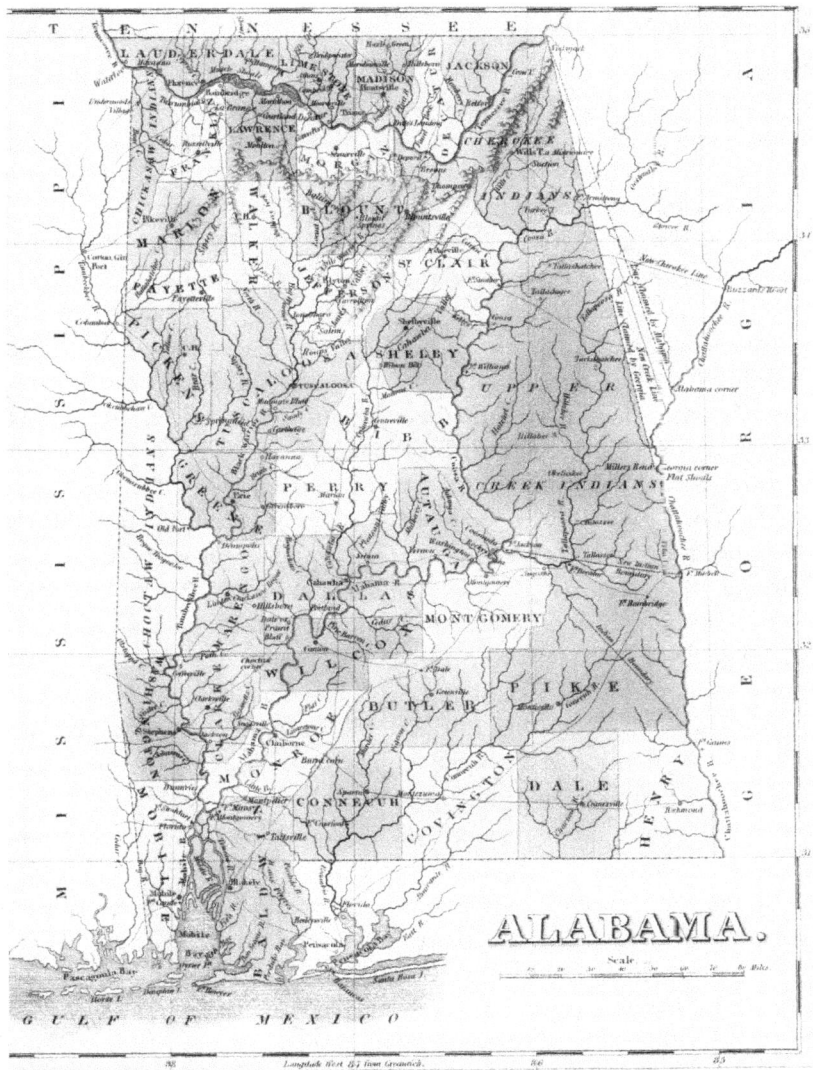

A map of the 1829 Old Creek Nation and Cherokee territory in east-central Tallapoosa County. *Courtesy of Peggy Jackson Walls.*

for $100; it was this sight [sic] that produced the "specimen of pure virgin gold" that Wyatt took to the Mobile Register's *office in the spring of 1831.... The U.S. mint in Philadelphia received its first shipment of gold from Alabama—thirty-one ounces of gold amalgam and dust which the mint assigned a coinage value of $559.*[16]

Alabama Gold

In the early 1830s, the story of a Mr. Marable, who "cleaned" eleven and three-quarters pounds of pure gold in one day, brought a flood of miners to Randolph County. Marable's recovery of gold was followed with a great celebration that ended in a "free" fight. He was killed in the affray and never had the opportunity to enjoy his newly found wealth. A heavy rain flooded the area the next day, covering Marable's discovery. Although prospectors searched for years, the site wasn't located until 1895. The story illustrates the vulnerability of the lone miner working in the Old Southwest frontier, where a man could make a fortune and lose it, along with his life, all in one day.

By 1835, Alabama's first authentic gold rush was underway, as the early gold mining towns of Arbacoochee and nearby Chulafinee and Pinetucky attracted hordes of miners and their families. Overnight, Arbacoochee grew from a small village into a bustling, gold mining camp. A boon to the frontier town, the mining industry offered employment to about six hundred men. By 1845, Arbacoochee was home to about five thousand people. The disorderly, mining camp that sprang up around Arbacoochee consisted of mining supply stores, gambling establishments, hotels, saloons and brothels. Mining companies paid between $0.75 and $1.75 per day during peak production in the mid-1840s. Miners used picks to dislodge rock inside narrow tunnels and shafts and filled wheelbarrows with ore to be transported outside, where they crushed the ore with stamp mills. A process known as amalgamation was employed to draw gold from the crushed ore, after which the amalgam was heated to four hundred degrees and used to separate gold from the mercury. In 1832, almost three hundred years after Hernando de Soto and his men tramped through Creek land east of the Coosa River, enthusiastic observations were made by the *Times of London* and other papers, reporting, "Alabama gold…deposits yield from one and a half to two and a half pennyweights a day…and under the management of persons of sufficient enterprise and skill, would develop an immense fund of wealth."[17] The success of Alabama's gold mining industry depended on the lucky turn of a spade in a sandy creek, "gold diggings" from a hillside or the accidental discovery of an unusual rock gleaming from a creek bed.

Goldville

The discovery of the Log Pit Vein in northeast Tallapoosa County was an accident. Numerous times, prospectors walked by what appeared to be an

The Old Southwest

Documentation of James C. Johnson's purchase of Hog Mountain land on June 4, 1842, three years after his discovery of gold. *Courtesy of Tallapoosa County Probate Records, Dadeville, Alabama.*

old, rotten log until one day, a curious prospector decided to inspect the log closer. Breaking off a piece, he observed gold in what turned out to be a large quartz outcropping. With the spread of the story, the population of the piney back woods area quickly grew from a few scattered cabins into a gold mining town of 3,000 to 3,500 people mostly living in tents close to their digging sites. More permanent structures were built with pine or oak cut from the surrounding wilderness. They included fourteen stores, two hotels, several saloons, gambling establishments and a pit for cock fighting. Gold mining was the hub of the town's economy and stimulated a bustling commerce in the town. On January 25, 1843, the town was incorporated under the name Goldville. A number of other mines existed around Goldville. Birdsong Pits was the first and was owned by Edward Birdson and operated with the help of slaves between 1840 and 1850. Other mines included the Mahan, Ealy and Jones Pits. The amount of gold taken from Log Pit reportedly was $30,000.[18]

Michael Tuomey, Alabama's first state geologist, recorded the following description of Goldville in the first state geological survey in 1857: "The Goldville mine was discovered in 1842. The most productive portion of it was worked in the 'Log Pit,' where the richest part of the vein varied in thickness from 4 to 2 feet. It was worked at this pit to a depth of 105 feet. The ore was valued at $2 per bushel."[19]

Hog Mountain

In 1839, a Mr. Johnson discovered gold at Hog Mountain. Working with primitive tools, he dislodged ore and hauled it in an ox-drawn wagon to Hillabee Creek, where he washed the gold. In 1842, the original land title was made to James C. Johnson, marking the dimensions as #1/2NW1/2sec.15,T.24N.,R.22E. That same year, a ten-stamp mill and amalgamation plates were employed in the removal of ore from Hog Mountain, also known as the Hillabee mine and the Hogback mine, location Secs. 10 and 15 T. 24 N., R. 22 E.[20]

2

Tallapoosa County's Gold Mining Districts

Devil's Backbone, Eagle Creek, Goldville and Hog Mountain

Tallapoosa County was created by the Alabama legislature on December 18, 1832, with land acquired on March 24 of that year from the Creek Cession of 1832. Located in the east-central part of the state, the county is bordered by Clay, Randolph, Chambers, Lee, Macon, Elmore and Coosa Counties. The four gold mining districts in Tallapoosa County are the Devil's Backbone, Eagle Creek, Goldville and Hog Mountain. The Devil's Backbone District in the Lake Martin area had mining activity in the 1840s, from the 1890s until World War I and renewed activity in the 1930s. The Eagle Creek District is located between Goldville and Dadeville, the county seat since 1838, when Creek Indian removal was taking place. Eagle Creek was the site of numerous gold mining operations beginning in the late 1830s: Tapley, Griffin, Morgan, Jennings, Greer, Johnson and Hammock mines. Placer gold was panned in the streams and tributaries. No one gold discovery distinguishes the area, but it was actively worked by prospectors during the 1800s and the early 1900s.

Goldville, in the northeast corner of Tallapoosa County, was the scene of Alabama's second gold rush. Following the discovery of gold in 1842, gold diggers rushed into the sparsely populated piney backwoods, creating a population of about three thousand people overnight. Located three miles from Hog Mountain District, the Goldville District is about fourteen miles long. Hog Mountain is unique in the quantity and quality of gold in blue

ALABAMA GOLD

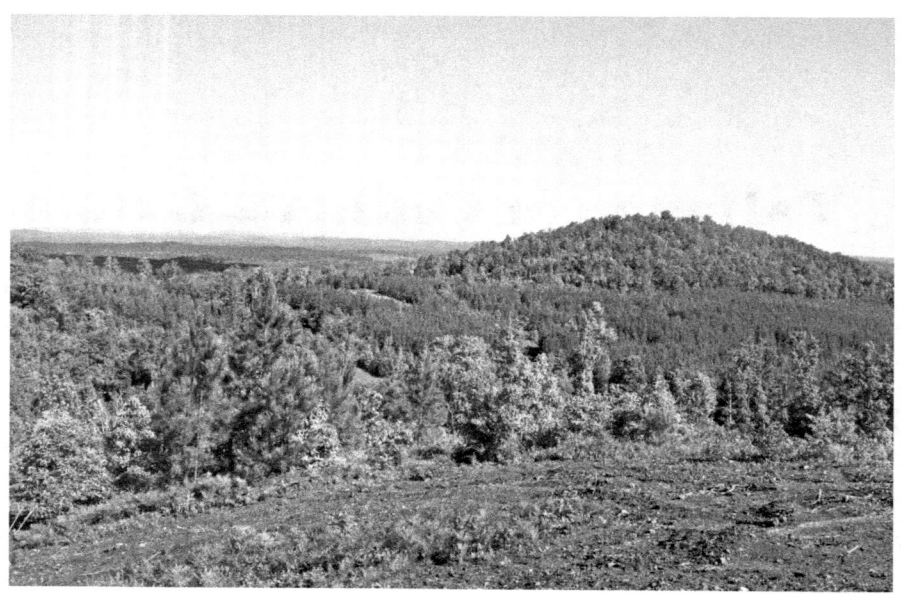

Hog Mountain ore was mined successfully in 1844 and 1845, using the crudest appliances and hauling the ore to the creek with oxen. *Courtesy of Peggy Jackson Walls.*

quartz veins of the Hackneyville Schist. The mine was the scene of three major gold mining operations: gold rush days, 1840–49; pre–World War I, 1890–1916; and the Depression era, 1933–37. The cyanide process was first introduced into the state at the Hog Mountain mine in 1903 by the Hillabee Gold Mining Company. Covered with pine, oak and hickory trees, Hog Mountain is situated near the junction of Hillabee and Enitachopco Creeks. Moore Creek drains the western side of Hog Mountain, and a tributary to Jones Creek drains the eastern side. Hog Mountain is also known as the Hillabee or Hogback mine.

Colonel B.L. Dean, businessman and mayor of Alexander City, wrote the following report regarding northeast Tallapoosa gold mines for W.B. Phillips, who included the notes in his 1892 Alabama geological survey:

> *The first work in this part of Tallapoosa County was done between 1840 and 1850 by Edward Birdsong….He owned and mined part of S.W. ¼ and N.W. ¼ of Section 4, T. 24, R. 23. His widow…could give more information about the mining interest in those days than any one I know. She said to me once that she was the cause of her husband's stopping work; the country was full of miners and she could not afford to raise her children*

The Old Southwest

where the Sabbath was a day of hunting and gambling. Her husband's work was carried on with Negroes. In illustration of the gold fever she said that her Negro cook, after attending to all of her duties at the house, would take her pan and wash out 75 cents worth of gold in a day, crushing the ore in a little hand mortar.

Towards the southwest we come next to the Jones Pit, in Sec. 5, T. 24, R. 23. On this property a great deal of work has been done with wooden stamps and the Arastra. There was also at one time a steam engine at the mine. Reports as to the yield of gold vary. The veins are from ten to twenty feet in width.

Next are the Ealy Pits in the S.W. ¼ of Sec. 26, T. 24, R. 22. A great deal of work was done here by Mr. A. Ealy and the Hon. Daniel Crawford, Ex-State Treasurer...He said that he made the machinery himself; four iron-shod wooden stamps run by water power at Jarvis Mill. He hauled the ore to mines, crushed it with the wooden stamps and then "rocked" it in a rocker....The best run he had ever made in one day...he said was $73 or $75. After the death of Mr. Ealy work was suspended, probably in 1845 or 1846, and has not been resumed since.

Heavy sulphurets begin to show at the Mahan Pits, but the ore carries also free gold...the ore...assayed $22 per ton.

Lastly, we come to the Ulrich Pits on the east bank of Hillabee Creek... Dr. Ulrich sunk several costly shafts, hunting for copper, but...finally discovered gold instead of copper, and erected a mill furnished with wooden stamps, taking the water for his power from Hillabee Creek. He worked in this way until the war [Civil War], making his gold into bars and buying cattle with it, so I am informed by old citizens...Ulrich's operations were conducted without the least regard to economical mining, and no thought for the future. Col. A.H. Moore [Hog Mountain] had some of the Ulrich ore assayed in North Carolina and told me that it ran $21 per ton.

On the west side of Hillabee Creek these seams continue in a south west direction, crossing the road from Alexander City to Hillabee Bridge on the Duncan Place. This belt seems to be bounded on the east by a large slate dyke from 200 to 400 yards distant from the quartz seams...This ends what I have to say about the Goldville Belt.

About two and a half miles west of the [Log Pit] we find a great mass of ore in the Hog Mountain. There are millions of tons of quartz in the Hog Mountain, all of it carrying gold. I saw assays of ore taken from 16 different places and they showed the ore to be worth from $4 to $16 per ton.[21]

Alabama Gold

Line of Gold Mines

The hillsides of Northeast Tallapoosa County are marked by an almost unbroken line of gold mines: Mahan Pits, Croft Pits, Stone Pits, Ealy Pits, Log Pit, Houston Pits, Goldville Pits, Germany Pits, Jones Pits, Ulrich Pits and Birdsong Pits. The mines start at the Duncan community, about six miles east of Alexander City, and extend through the northeast corner of Tallapoosa County into Clay County.

One of the most famous mines in northeast Tallapoosa County is the Dutch Bend, or Ulrich, due, in part, to the popular story of how gold was discovered at the site. Dr. Ulrich was a native of Germany who came to northeast Tallapoosa County in 1840 with a group of settlers from Savannah, Georgia, looking for a terrain similar to his homeland, where he could settle, plant vineyards and produce commercial wines. He purchased 1,200 acres of land near Hillabee Creek. In the process of digging a large tunnel into the hillside to use as a wine cellar, he discovered a vein of gold. After finding other veins of oxidized ore, he built a stamp mill to process the ore, powered with water from Hillabee Creek. Thinking Ulrich was Dutch, locals referred to the property as Dutch Bend because it was located on a sharp bend in the Hillabee Creek. Dr. Ulrich hauled the ore from the tunnel in small cars on wooden tracks. He kept no records as to cost and production; therefore, it is impossible to know how much gold was produced at Dutch Bend in the early years of gold mining in Alabama. Dr. Ulrich had the gold made into small, one-ounce bars and traded them for mining supplies and other purchases.[22]

> *Many old workings, some of them tunnels, from 50 to 700 feet all driven in the immediate vicinity, show the crude mining work of half a century ago. It was here Dr. Ulrich extracted ores by his slave labor, but in no instance did his work extend to greater depth than 20 to 30 feet. He only sought the decomposed and free milling ore. All the work done clearly established the Geological features of the formation. The veins are "True fissure," extending vertically down to unknown depths and increasing in width with downward development.*[23]

A great deal of alluvial mining was done in Hillabee Creek and other branches in northeast Tallapoosa County. One of the first methods pioneer miners used to retrieve gold from alluvial deposits was hand panning. The miner filled his pan with one-half alluvial sand and one-half gravel. Adding

The Old Southwest

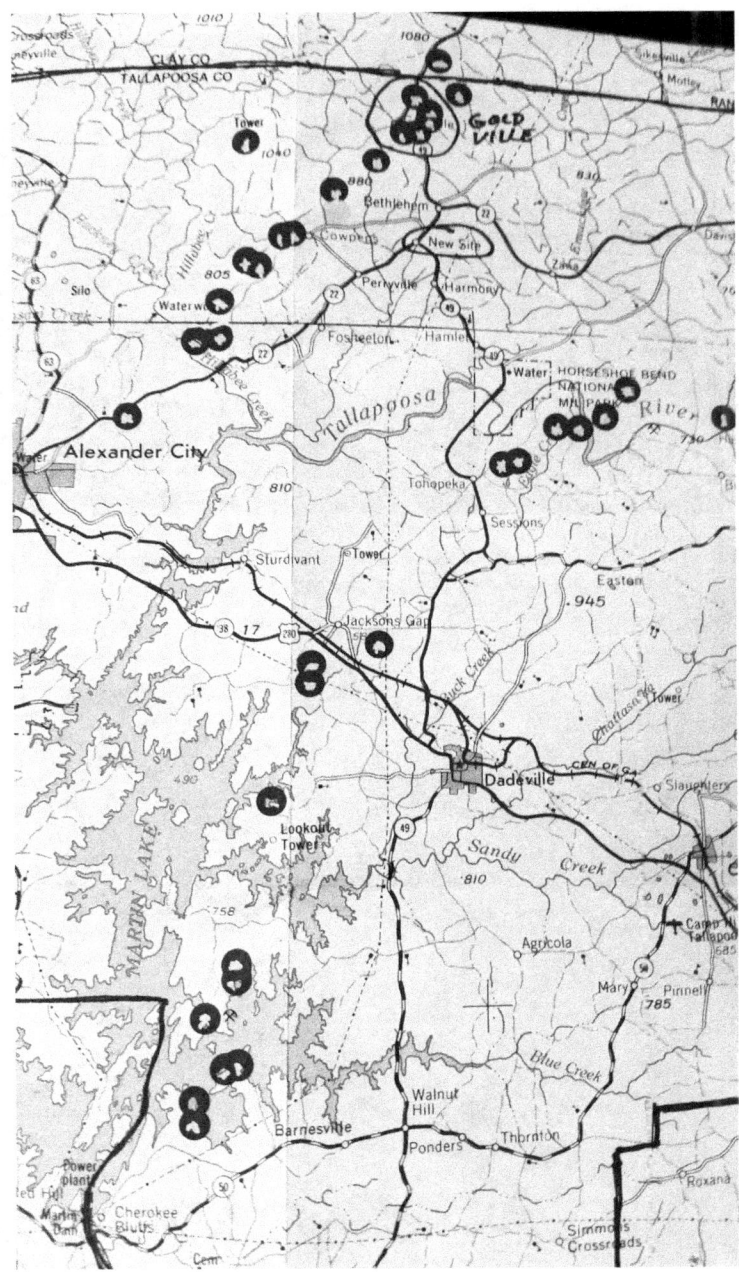

A line of gold mines extended from Duncan, five miles outside Alexander City, to northeast Tallapoosa County. *From Big Ten Maps; courtesy of Tallapoosee County Historical Museum.*

Alabama Gold

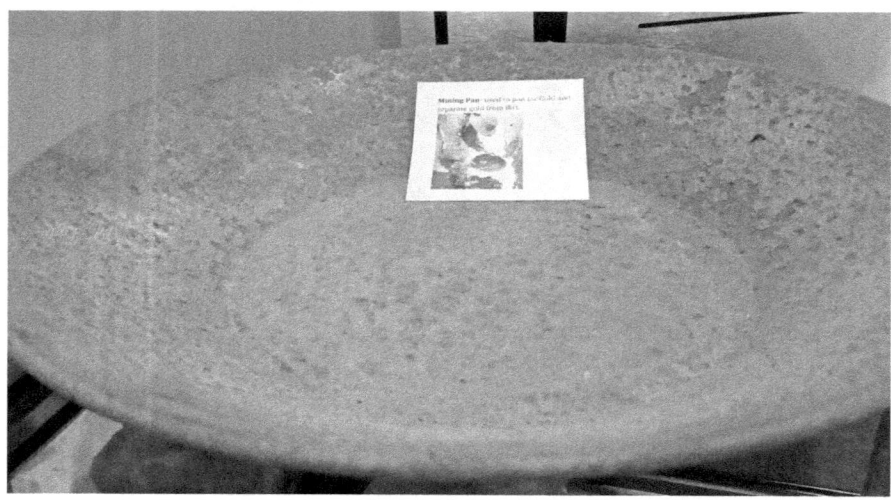

This picture of a gold mining pan is used with permission of Pine Mountain Gold Museum, located in Villa Rica, Georgia. *Courtesy of Peggy Jackson Walls.*

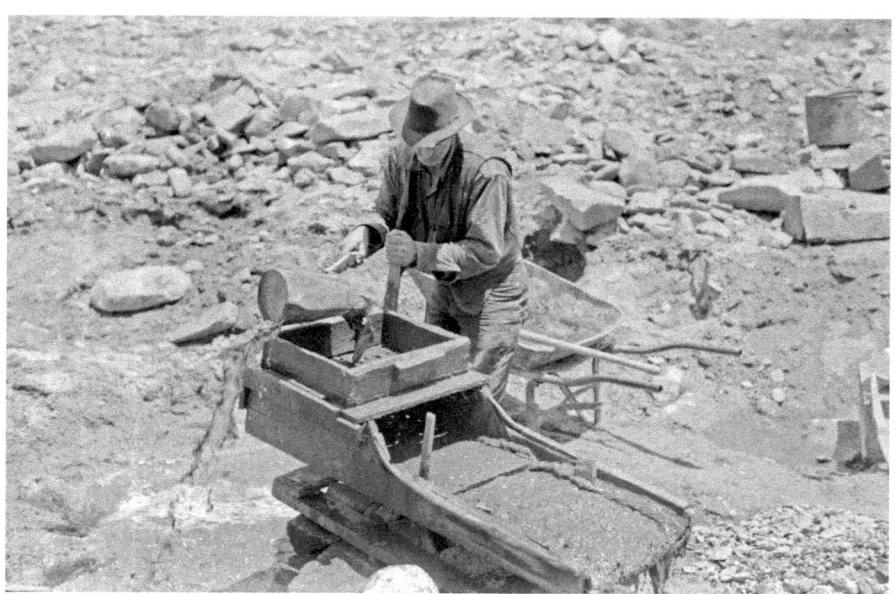

The miner is using a sluice box to recover gold from the mixture of water and ore. *Courtesy of Library of Congress.*

The Old Southwest

water to the top, he shook the pan vigorously side to side at an angle to separate the gold from the rock. Other devices included the gold mining rocker box, a wooden trough with deep cross riffles resembling a washboard. The trough was mounted on rockers. The miner would feed water and alluvial material into the upper end. The rocking motion moved the sand and gravel through and out of the trough, leaving the gold in the riffles.

The old-fashioned sluice box had removable troves that were about one foot wide and often hundreds of feet long to accommodate large amounts of gravel that twenty or thirty men could shovel into it continually as a large stream of swift water washed over the ore. The final step was to remove gold from the box.[24]

Southern Tallapoosa County Gold Mines

The Devil's Backbone District in the Weodowee formation extends west of the Tallapoosa River as far as Elmore County. On the east of the river, the district continues in a northeast direction into Chambers County. Rocks in the belt are primarily slate, phyllite, quartzite and schist of sedimentary origin. In the middle section, the beds dip at steep angles toward the southeast with heavy sections of quartzites exposed in a ridge, known by locals as the Devil's Backbone. Antebellum miners worked in the auriferous gravels along the streams without gaining any large amounts of gold. Many of the mines now are covered by Lake Martin.[25]

Blue Hill Gold Mine

Jonathan Steiner on March 11, 1843, acquired a patent from the US government for the land that would later become known as Blue Hill. By 1845, James Dowd Phillips had purchased three hundred acres, called the "Winn Creek Place," the site also of a Winn Creek post office. James Dowd Phillips and his wife, Sarah Ann Hampton McNiell, immigrated to Dadeville in 1845, following other settlers and prospectors from Georgia. With the gold rush at Goldville in northeast Tallapoosa County and a town of around three thousand people, "gold fever" was rampant in Tallapoosa County. In 1849, when the cry of "gold" came out of California, thousands of miners abandoned their "diggings" and headed west with the hope of striking it rich in the western gold fields. Among them was forty-four-year-old James Dowd Phillips, who left behind his thirty-five-year-old wife, Sarah Ann, to take care of the family estate with the assistance of servants. James and Sarah had five girls ranging in age from two to nine, with a child on the way. Born while his father was prospecting in California, the son was named Charles California Phillips. James continued mining in California until late 1851 or early 1852. After he returned to Tallapoosa County, another son, Josiah Samuel, was born on November 14, 1852. Sarah Ann saved the letters James wrote to her while he was in California. The following, postmarked "California Mines Nov. 10th/50 near Mocalumne River," documents the date and location of the mining camp where he worked.

A Letter to Sarah Ann Phillips

In the following letter to Sarah Ann, James described life in a mining camp located near Mocalumne River.

Postmarked: Sacramento, Cal.
Mrs. James D. Phillips
Winn Creek P. Office
Tallapoosa County, Alabama

California Mines Nov 10th/50 (near Mocalumne River)

My Dear Sarah Ann

I have again delayed writing you much longer than I should only that I have been anxiously waiting a letter from you yet I have not rec'd any since I wrote you; and having now an opportunity of sending this to the office I write you a short letter and will write you more as soon as I get your letters—which I hope will be in a few days as I hear there are some at the office for me.—I have just got out of a sharp spell of fever & chill & fever. I was taken about the last of Sept. with the Billious fever at Sacramento City to which place I had gone with my team to haul provisions to the mines. I hired a man to drive my team back to the mines and I rode on the wagon; I could not bear the idea of being sick long in S. City as it is indeed a place of very bad water and much filth about the streets. I got back to my cabin at the mines quite sick—but in a few days broke the fever. I then took the chills & fever which held me for some time so that I am now just getting cleverly about and feel quite well only that I am yet weak—you will I know sympathize much with me for my sickness and I fear you will conclude my health getting worse instead of better—but I am clearly of the opinion that my general health is much improved—especially my old complaint as I have not had any trouble from cough since last spring and very little sick head ache to which I had been so subject. The exposure of driving a wagon sixty miles of the most dusty road you ever saw and most of the way very bad water—and then the bad water and filth of a City worse if possible in warm weather than Mobile or N. Orleans was enough to give much stouter constitutions than mine sickness.—The water where we are in the mines is good and there has been but little sickness about us—the weather is now very pleasant—with frosty mornings—and the season for chills & fever proper. There has been much sickness this summer in some of the Northern mines also on the Sacramento River & in S. City—it has mostly been among those who came over land this year—and many of them have gone to their last homes—the immigration has generally been very sickly this year (last year generally healthy) and the number of deaths before they reached California is estimated as high as ten thousand and hundreds have died since they got in. I have scarcely heard of a death among those who have been in the mines all summer—Sacramento City has been for some 4 or 5 weeks the scene death & destruction from that fill destroyer the Cholera—it has doubtless been very bad probably worse according to the population than ever it was in N. Orleans—it is said not to be so bad but seems now to be abating—there seems to be but little or no apprehension of it in the mines—not now that we of Tallapoosa County feel it is raging in Mobile & N. Orleans.

As to the news of the success of the miners generally this summer it has been very poor indeed hundreds of them have not made a cent but spent all they had made before. The rivers have not fallen so low this summer as they formerly had and they have not proved rich only in few places in the mines are not more than able to buy provisions for winter—and many of them not that—yet, some few made their pile in a few days and go home but this number is scarce and far between—The general hope now is for good winter digging—and great quantities of provisions are stored up in the mines and many in the cabins for winter quarters—provisions are from 25 to 50 cents pd & everything sells by the pound, not measure. [Margin Note: *Write often and fully without fail. Direct your letters to Sacramento*

City as I believe I can get them more conveniently from there. I am providing much more comfortable winter quarters than we had last winter. We have had no rain here yet.]

My dear wife. I know you are anxiously waiting for me to say when I am coming home but you are not more anxious than I am yet I cannot now set any time precisely—if I can wind up my business here profitably I shall be at the pleasant little home of my wife & little ones next spring—if my health should not be good I shall be sure to return this winter or spring— should my health seem to improve and I hope it will and my prospects here are anyways flattering I may stay longer.

I flatter myself that you have fortitude and perseverance to bear up and cheerfully glide along during our long absence—may Heaven in all kindness protect & guide you, give you comfort in all your trials is the best that I can wish you—I rec'd a letter from brother a few weeks ago. I was glad to hear that he had called to see you—he says you are getting on well and he thinks will do well for some time to come—which I hope is the case I feel great anxiety for you all—I hope our dear ones are learning their books and all that is good and now my dear little ones be obedient to your Ma learn to love your books be industrious and helping domestick [sic] affairs. Learn to love and fear God—and you will have the love and esteem of all good people and it will afford your Pa so much pleasure on his return home to find you—growing up in industry virtue truth and literary accomplishments—your Pa often thinks of you and your dear Ma and Prays that God may protect you all.

William and Charley are both well and send much love to you all—we are making some money but not a fortune in a few days by no means—we hope to do better this winter.—My best wishes to John & Livona S. C. McG & Mary & little ones with all inquiring friends—my best Wishes to each servant—I hope they are good & obt to you and that they will continue so—howdy to all—May heavens choicest blessing attend you all is the most earnest wish & prayer of your ever affectionate husband.

J.D. Phillips

[Margin Note: *I am providing much more comfortable winter quarters than we had last winter. We have had no rain here yet. Excuse any mistakes as the bearer is ready to start & I have not time to correct. Think not that our long separation is for the worse but I hope all will yet be for the better. I know that you can very readily appreciate the reasons that induce me to stay* here to make something—and I believe and trust that you will persevere and keep all things in as proper order at home as your circumstances will admit. Kiss all the little ones for me 10,000 to yourself.]²⁶

History of Blue Hill by Ben E. Kidd III

Alabama gold miners like James D. Phillips and others who struck out from the southern Piedmont to make their fortune in the West were leaders in the California gold mining fields. They were experienced in the different methods of mining and surviving in a hostile environment. But the trip west was long and fraught with dangers. Ben E. Kidd III in his summary of Phillip's trip to California referred to H.W. Brands's *The Age of Gold* to describe his trip and means of travel.

> *The most used initial travel route to California for Americans in 1849 was by ship. If the route chosen was a sea voyage around the tip of South America, it required five or six months, depending on the conditions off Cape Horn. The closest and fastest route was by ship to the Central American isthmus in Panama at Chagres, the port on the Caribbean side of Panama. The land crossing was over sixty or so miles of mountains and jungle between Chargres and Panama City on the Pacific side. The first leg of the land crossover was usually made by ascending the Chargres river to Gorgona, a distance of about fifty winding miles using natives in their dugouts. Paddles were used to row upstream until the elevation and rapid current required poles and great effort to push the craft...the final leg taken to reach San Francisco was to catch a ship from Panama City. By late 1849 and during 1850 normal departures were problematic because many ships lost their entire crews who jumped ship in San Francisco to seek their fortunes in the gold fields. Hundreds of ships were left stranded in the harbor for lack of crews to man them.*
>
> *James D. Phillips' letter to his wife dated 10 November, 1850, indicates that he passed through Mobile and New Orleans on his trip to California. It is most probable that he chose the isthmus route and if so, he could have departed on the ship Crescent City of the Pacific Mail steamship line. Their price of ocean passage from New Orleans to Chargres and from Panama City to San Francisco inclusive ranged from $200 to $500, depending on the class of shipboard accommodations.*

Alabama Gold

James D. (Dowd) Phillips' letter of 1850 provides insight into the man himself and his surroundings. He was obviously well educated, pious, and diligent and loved his wife and children deeply. He also wanted to provide for the wellbeing of the two unidentified men from home who accompanied him to the California Mine to work with him. His California mine venture was apparently profitable. He died in 1866 leaving his estate to his wife. His holdings included the 300 acre Winn Creek Place, a 600 acre plantation, land in Clay County and a town home with several adjoining lots in Dadeville on Lafayette Street.[27]

In the Devil's Backbone District, Blue Hill, Gregory Hill and Silver Hill were all located on the same vein system, three miles long and in places one hundred feet wide. In the early days of gold mining, a large quantity of alluvial gold was retrieved from Blue Creek, which ran parallel to the large vein. The ore was of a fairly low grade and contained graphite, which caused mercury to slide off the plates of the stamp mill used at the site. By 1850, gold mining activity had subsided. Large numbers of miners headed west to join in the western gold rush taking with them Alabama gold for a grub stake and mining experience that would give them an edge against the novices coming from all over the United States. In Tallapoosa County, mining was left to a few men who were adept at handling both a plow and a gold pan.

Historian Otis Young remarked that the $40 million gold production of the South before 1861 appears relatively miniscule when compared to the $550 to $680 million production estimated for California between 1848 and 1861. However, the effect of the southern gold mining industry on local economies was significant, encouraging speculation and investment in the mining industry.[28]

3
"It's Good to Be Shifty in a New Country"

ROBERT GRIERSON

The Creeks, also known as Muscogee, were one of the Five Civilized Tribes who had adapted to the European lifestyle. The more prosperous Indians owned black slaves. Promoting assimilation of the Indians into the white culture, the federal government appointed Indian agents to live among the Indians. Benjamin Hawkins was the Indian agent for the southeast territory and friends with the Robert Grierson family, whom he described as "the most prosperous of the trader families in the Upper Creek Nation." As a licensed trader with the Creek Indians, Grierson was respected by Indians and by US government representatives, interacting with both on a regular basis. His wife, Sinnuggee, hired women from the Hillabee villages to pick cotton and to work in the fields. Women were paid with a half pint of salt or three strands of small glass beads for each basketful of cotton they picked or a half pint of rum after they picked two basketsful.

In "The Letters of Benjamin Hawkins," a collection of personal correspondence, Hawkins referred to Grierson's involvement in the Revolutionary War, stating, "Inasmuch as there is no other reference in Alabama history to any other white man living in what was subsequently the State of Alabama than Mr. Grierson, he should be credited with being Alabama's only Revolutionary soldier."[29]

Grierson's prominence is also documented in the "Memoirs" of Marinus Willet, a member of Washington's staff. Willet was sent by President George

Washington to invite Alexander McGillivray and the Creek chiefs to the 1790 New York conference and located him "in the Hillibees" at "the hospitable mansion of Mr. Grierson."[30] As conditions worsened for the Creeks and for the Grierson, now Grayson, family, Robert Grierson appealed to Indian agent Benjamin Hawkins:

> *We are injured in our property, we are told to go to the protection of the Alabama laws—to present our case before an Alabama court. We present our case, and we are not permitted to be heard in behalf of each other. Our cause is adjudged by a jury of Alabama, under the direction of a court of Alabama, administering the law of Alabama. The law, if it contains a single provision which can protect the Indian from outrage, or can redress his wrongs when they have been sustained, is, to this extent, unknown to us. We know it only an instrument by which we are oppressed, and as opposing an insurmountable obstacle against our obtaining redress.*[31]

After the defeat of the Red Sticks and destruction of the Creek Nation, Robert Grierson and his Creek family remained in the area, where he died in 1825 of natural causes. In the 1830s, his descendants suffered through the displacement of Indians by settlers and speculators who rushed in to claim a piece of America's new frontier. Descendants of Grierson became known as the Graysons, and some, such as George Washington Grayson, became famous leaders in the federation of tribes in the West.

GRIERSON DESCENDANTS

A direct descendant of Robert Grierson, George Washington Grayson was an officer in the Second Regiment of Creek Confederate Volunteers and commanded Company K. He was for a time treasurer of the Creek Nation of the West, secretary of the five tribes and principal chief of the nation and represented the Indians as a delegate in Congress prior to the admission of Oklahoma.[32]

Robert Grierson was honored when a Daughters of the American Revolution chapter was organized at Headland, Alabama, on November 12, 1947. The DAR chapter was named Robert Grierson for the only man from the present state of Alabama who fought in the Revolutionary War on the side of the colonists. The irony is many of his descendants were removed from the land he fought to save for American families.

George Washington Grayson wrote a four-volume autobiography that was the basis for Claudio Saunt's *Black, White, and Indian: Race and the Unmaking of an American Family*, providing for a modern audience a view of the Grayson family, Native Americans and African Americans during the influx of white settlers, illegal land transactions, slavery and the Trail of Tears.

ALABAMA CENSUS

In 1830, the state population had increased from 127,901 a decade earlier to 309,527 and then to 590,756 in 1840, reflecting the numbers of settlers, miners, entrepreneurs and speculators arriving during the Alabama Fever.

Claudio Saunt notes in *Black, White, and Indian*, "Thousands of whites moved onto their [Creek] lands, notching trees to mark their claims, sometimes in the middle of cornfields actively cultivated by Creeks" and names Henry Towns and Mr. Rhoden as two offenders. By the end of 1831, the Creeks were desperate. George Washington Grayson appealed to President Andrew Jackson for help, describing the harsh conditions in which the Native Americans lived: "Your White children are fast settling up my country. They are building houses, Mills…and destroying all my timber and games." Neha Micco, a prominent Creek chief, confessed, "We expect to be driven from our homes."[33] In this hostile environment, the Indians had no protection from the military troops, nor did they possess legal recourse when they were threatened or their property was stolen or damaged. Creek leaders signed the final treaty with the US government, the Treaty of Cusseta, on March 4, 1832, in which "the Creek tribe of Indians cede to the United States all their land, East of the Mississippi river."[34]

In 1835, as the Creeks lost their land and legal rights, Alabama's first gold rush was in full force at Arbachoochee in Cleburne County. George Washington Grayson described the culture of lawlessness in central Alabama during gold fever and land rush days:

> *Theft reached its greatest proportions in the first three months of 1835, just before charges of corruption led the government to suspend all sales. One might expect the Graysons to have survived the onslaught, like most Creeks, quietly and desperately. A few Creeks did not survive at all. They starved to death, died in drunken brawls, or were shot down by white intruders.*[35]

Alabama Gold

Fraud in Land Dealings

Land speculators took advantage of the Creeks' situation by illegally purchasing Creek allotments and selling them to settlers. Saunt observed, "Other strikers dispensed with legal ruses and simply ran the rightful residents off their land and set fire to their houses." Hostilities between whites and Indians resulted in a final conflict in 1835, after which "U.S. troops, assisted by Georgia and Alabama militia and led by General Winfield Scott, forcibly rounded up Creeks and sent them to Indian Territory. Some went in chains, under the watch of armed soldiers. Creeks had to begin life anew in lands west of the Mississippi."[36]

Creek leader Opothleyaola described how unscrupulous speculators profited from the sale of land allotted to the Creek families by the US government: "A 'fiendish designing scoundrel' would hire an Indian to impersonate the owner of an allotment. After certification of the sale, the impersonator returned the money to the purchaser, saved five or ten dollars, 'given to the Indian as a premium for his rascality.'" Opothleyahola concluded that in this way "a few hundred dollars and four or five Indians could sell all the land in the Creek purchase....In the other common method of swindling, strikers would purchase land from the rightful owner. By force or fraud, they later recovered the payment from the seller. Said one striker, the 'best of it was' that this method allowed him to recoup his entire outlay."[37]

Johnson Jones Hooper: The Old Southwest Humorist (1815–1861)

No one was more familiar with the rugged elements of the Old Southwest—the tall tales, dishonest land transactions and frontier humor—than humorist Johnson Jones Hooper. In 1835, Hooper moved to Dadeville from Wilmington, North Carolina, during Alabama's "flush days." He was attracted by the adventure of living on the Alabama frontier and building a prosperous life in the Old Southwest. Based on his experiences as Tallapoosa County's first census taker, Hooper wrote the satirical essay "Taking the Census in Alabama," which was published in 1843 in the *East Alabamian* newspaper and later reprinted in William T. Porter's *Spirit of the Times*. As editor of the *East Alabamian* newspaper during the gold rush years of 1842–45, Hooper heard stories of and perhaps witnessed deals made between land speculators and Creeks, who were being pushed out of their homeland.

From Hooper's book *The Adventures of Captain Simon Suggs*, the following story demonstrates how Suggs made every business deal "work to his good" and fulfill his philosophy that "it's good to be shifty in a new country." The opening lines of Chapter Six set the tone for the manner in which Indians were treated: "There are few of the old settlers of the Creek territory in Alabama, who do not recollect the great Indian Council held at Dudley's store, in Tallapoosa County, in September of the year 1835. In those days, an occasion the sort drew together white man and Indian from all quarters of the 'nation'—the one to cheat, the other to be cheated."

The story focuses on speculators who wish to gain valuable land owned by an Indian known as "Sky Chief," who adamantly refuses to sell the land. A speculator named Eggleston gains control of the land through courting and marrying Sky Chief's fifteen-year-old daughter Litka. Eggleston promises to take care of Litka and her father when the rest of the Indians are forced to leave; they can stay with him "by the graves of their fathers."[38] Sky Chief decides this is "good talk" and signs away his land title to Eggleston, who sells the land for $3,000 and then refuses even to provide a wagon to help them during the Indian removal to the West. The story was an episode in Johnson's writing about Creek frauds.

Other episodes in Hooper's *Captain Simon Suggs* illustrate how frontier con men like Simon Suggs are eager to "best" other land speculators who are no more honest than he is. Such is the case in the tale of Widow Rugsby and the land she has promised to sell Captain Simon Suggs. Perhaps flattered by the captain's attention and his military title, the widow refuses to consider attractive offers from other speculators. Knowing the captain has no money, the other speculators persist in their offers to the widow. The captain disappears for several days after suggesting he is going to borrow money from a friend. He returned before the widow sells the property to one of the other speculators. Seeing two bulging saddle bags, the speculators, "reckoned" the captain had "struck it rich." They were eager to obtain the choice parcel of land and paid him more than "two and a half times" what he had paid the widow. With the transaction completed, the captain leaves, later emptying the saddlebags of rocks on the side of the road, congratulating himself for his good business sense and lack of greed.[39]

Alabama Gold

Tallapoosa County 1840 Census

In his role as narrator in "Taking the Census in Alabama," Johnson Jones Hooper describes Tallapoosa County as "thinly populated" in many places; the hill people were suspicious of census takers coming "to count the noses of the men, women, children, and chickens resident upon those nine hundred square miles of rough country which constitute the county of *TaHapoosa*." This job "took Hooper into every nook and comer [sic] of the hilly, muddy county and gave him ample opportunity to see Alabama frontier life at its crudest." Seeing Hooper approaching at a distance, children would run to warn their parents that "the chicken man was coming." Settlers feared the census man would count and tax every chicken, child and possession they might have. Some of the backwoods women refused to name their children and instead just gave them a number. When Hooper finished, the total count was "2,318 white males, 2,106 white females, 2,013 slaves and 9 free colored persons 6,444 in all."[40]

Captain Simon Suggs, Late of the Tallapoosa Volunteers

As editor of Lafayette's first newspaper, the *East Alabamian*, Hooper began writing Whig editorials and short stories. When he lived at the old Dennis Hotel in Dadeville, he; a brother, George; and Bird H. Young entertained themselves by swapping stories about the "backwoods" folk with whom they came in contact, exaggerating the dishonesty and ignorance of people to create colorful tales, characteristic of the Old Southwest humor.

Intrigued by the tales and his own experiences with settlers, speculators and clients, Hooper wrote in 1845 what is arguably the most influential work of fiction produced in the Old Southwest: *Some Adventures of Captain Simon Suggs, Late of the Tallapoosa Volunteers*. In the character of Simon Suggs, he created the stereotypical frontier con man, who appeared, slightly modified, in the character of King in Mark Twain's *Huckleberry Finn* and in William Faulkner's character Abner Snopes. In his notes, Johnson Jones Hooper reveals he based the character of Simon Suggs on the real Bird H. Young, who came to central Alabama a year after Jackson's Indian Removal Act was passed and Tallapoosa County was formed. Hooper followed in 1835. By 1840, the two had met since B.H. Young appears on Tallapoosa County's first census as number 137.

The Old Southwest

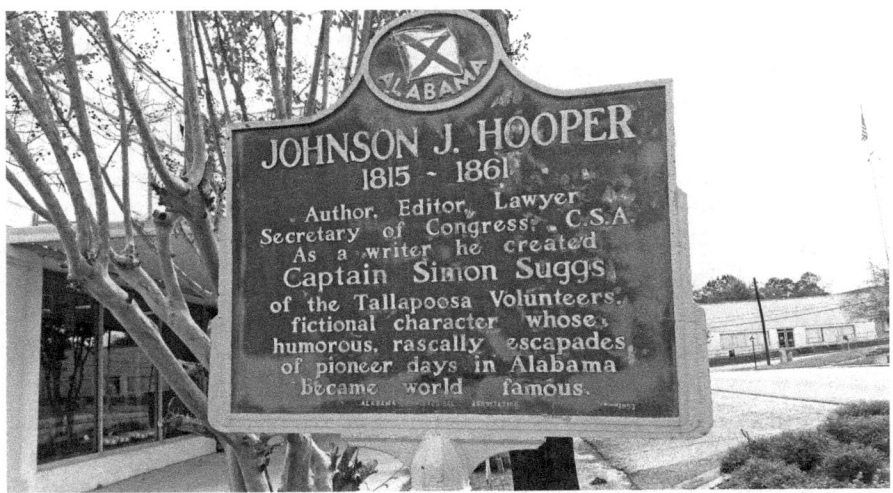

This plaque honors Johnson J. Hooper (1815–1861), author, editor, lawyer, creator of the fictional character Captain Simon Suggs of the Tallapoosa Volunteers. The plaque was donated by the Alabama Historical Association in 1953. *Courtesy of Peggy Jackson Walls.*

Although Johnson Jones Hooper was a lawyer, editor and secretary of the Congress of the Confederate States of America, he is best remembered for his creation of Simon Suggs, a frontier rascal whose fictional escapades were purported to have been based on the real life shenanigans of Bird Young, who was arrested in 1834 for gambling. In 1835, he was fined for gambling, and in 1836, he was arrested for "Betting at Faro." Through his frequent brushes with the law, Young became well known and popular in the frontier culture, where taking chances was respected and disregard for any type of restriction was common. The sharper a person's "mother wit," the more he was admired and promoted in local offices. Bird Young was Tallapoosa County's first tax collector, a justice of the peace, a constable and the guardian of several estates. The contrast between Young's responsible and respectable offices to which he was elected or appointed and his scandalous reputation attracted Hooper's attention, and his deeds could have been the basis for Captain Simon Suggs's adventures in Tallapoosa County as the following example suggests.

On one occasion Bird Young is reputed to have stopped at Coosa Hill, a Wetumpka hotel, and, just prior to leaving, sneaked his saddlebag and blankets out and hid them nearby. He then returned and politely asked the proprietor to assemble his belongings and prepare his horse for departure. After searching in vain for his guest's baggage, his innkeeper was forced to deduct twelve dollars from the bill, and Young, of course, promptly paid the reduced bill and as promptly retrieved his "lost" property.[41]

Both Bird Young and Simon Suggs were gamblers and speculators who were determined to make everything work to their advantage. The epitaph on Bird Young's headstone in the Dark Cemetery in Alexander City reads simply, "Bird Young, Alias Simmon Suggs, married Ann Donaldson, Born about 1800, Died about 1870."

Through shady land deals in which Indians were persuaded to forfeit their land, a number of crafty people became wealthy; however, the majority of settlers were hardworking farmers and gold diggers, eager to make their mark in the new frontier. They were often shrewd business people who managed to "get ahead honestly." The satire of Johnson Jones Hooper, a lawyer and editor, portrays life on the old southwestern frontier, where shrewdness and dishonesty ruled many business transactions and supported the philosophy of the fictional Suggs that it's "good to be shifty in a new country."

MICHAEL TUOMEY, ALABAMA'S FIRST STATE GEOLOGIST (1805-1857)

While editor of the Lafayette paper, Johnson Jones Hooper described settlers and events in Tallapoosa County, such as Indian footraces and land speculation deals. But he wrote at least one editorial criticizing a well-respected state official, Alabama's first state geologist, Michael Tuomey. He disagreed with Tuomey's evaluation of a Coosa County silver mine. Tuomey replied in his own defense:

> *A friend has directed my attention to a communication, copied from the* Montevallo Herald....*I hope you will allow me a little space, in which to say a word or two, in relation to my connection with this Mine.*
>
> *I am not altogether unaccustomed to attacks of this sort, although they do not often so imprudently appear in print—the very substance of the communication in question, has been repeated, over and over, at places of public resort—nor should I notice an anonymous paragraph, had it not derived importance from its appearance in your paper; you yourself, Mr. Editor* [Johnson Hooper], *in a kind and polite paragraph, in your paper of the same date, seem to intimate that it requires some attention.*

The Old Southwest

In his response, Tuomey defends his evaluation and his qualifications. His experience included an appointment as state geological surveyor in South Carolina in 1844, prior to moving to Alabama and accepting a faculty position in geology, mineralogy and agricultural chemistry at the University of Alabama when offered the position in 1848. One of his obligations was to create an annual geological survey of the state's mineral resources. He filled the position of Alabama's first state geologist for four years without pay and was forthright in his theories and criticisms of methodology and evaluation of Alabama gold and other minerals. His estimation of the gold mines in Alabama were capsuled in a report in the *Montgomery Advertiser and State Gazette* of January 3, 1855:

> *The interest excited by the gold mines of the Southern States is very much dependent on the fluctuations in the price of the all-absorbing staple, cotton. During depressions in the cotton market, ordinary labor is frequently driven to seek more profitable employment in the gold mines and particularly in those mines known as branch or deposit mines, where the nature of the work requires but little capital, and no great exertion of skill. The deposits are generally run over in the most careless manner, and without reference to economy in the working, or regard to the future value of the mine.... This has been the brief history of all the attempts made to explore the gold mines of the State.*[42]

In the *Cahawba Valley* newspaper of November 10, 1847, Michael Tuomey expressed his impatience with the gold hunters of his day:

> *Give me a genuine hunter for a guide. It is true he may know nothing of your metamorphic, your Silurian, or your Carboniferous rocks; but he knows grindstone grit, and lime rock...He knows where the precipice overtops the pine, and where the shelving rocks form the lair of the wolf; and although these localities may be ten miles distant, you are as certain to go to them in a "bee line" as his rifle would be to bring down a deer at 80 paces. But above all save me from your searchers after the precious metals, your gold hunters—knowing, mysterious men, that will waste you a whole blessed day to show you where they picked up something strange (ten to one a scale of mica or bit of pyrites) and then will be unable to find the place.*[43]

While acknowledging the presence of gold in Alabama, Tuomey makes his case for investing resources in mining "the homely products of coal, iron, and stone."

Alabama Gold

The truth is, when we read of the vast amount of precious metals yearly raised from any given mine, we are too apt to think that all this dazzling product is so much clear gain. It is forgotten that gold can no more be separated from the soil without labor, than coal or iron; and that the purification of the precious metals involves vast outlay for materials and machinery. Hundreds of the seekers for the glittering treasure, have involved themselves in irrecoverable ruin; while the humble smelter of iron has rarely failed steadily to advance in prosperity and wealth. There are many who are proud that Alabama is a gold-producing State; but, for our own part, we would rather that the gold were given to our neighbors; and that, in its place, our iron, already so abundant, were everywhere diffused. There are many who will receive this opinion with a smile; who will yet live to make it their own.[44]

Michael Tuomey's death in 1857, due to pneumonia and heart problems, ended the career of Alabama's first state geologist, whose studies and recommendations laid the foundation for future studies of geology in Alabama. His insight and careful analysis of the state's mineral resources led to a greater understanding of Alabama's mineral resources.

4

Cotton Boom and Bust, Lost Confederate Gold, New Interest in Gold Mining

Following the exodus of gold miners to California in 1849, interest in gold mining in the Southeast waned. Prospecting was left to farmers, who were more interested in "white gold" (cotton) than digging for the "yellow stuff." "In the decade before the Civil War cotton prices rose more than 50 percent, to 11.5 cents a pound" and "the US cotton crop nearly doubled, from 2.1 million bales in 1850 to 3.8 million bales ten years later."[45] In Alabama's Black Belt counties, such as Lowndes and Montgomery, cotton plants flourished in the rich, fertile soil of the region. Slave labor was used to plant, tend and pick the cotton for shipping to northern textile mills and to European manufacturers. The rocky terrain and poor soil in Tallapoosa and surrounding counties offered little incentive for large-scale cotton farming.

Trading the Pick and the Plow for the Musket

The heart of Dixie did not beat so strongly for secession as people might imagine. When debate about state secession arose, many Tallapoosa citizens viewed the "Battle between the States" as "a rich man's war and a poor man's fight" and were opposed to the state's seceding from the Union. The

Old Southwest humorist, lawyer and politician Johnson Jones Hooper, however, was a strong advocate for immediate secession.

At the Secession Convention, county delegates voted fifty-four for secession and forty-six against. Although Tallapoosa County voted against passage of the "Ordinance of Secession" on January 7, 1861, when the ordinance was passed, Tallapoosa County began to prepare for war with the other counties. The Hillabee Blues unit was organized in northeast Tallapoosa County and held muster in the fields near Hackneyville. When the Provisional Congress of the Confederate States of America convened in Montgomery, Hooper was elected secretary of the Congress. In 1862, he left the position and served as editor of the Congress and the Confederate States of America.[46] Although Tallapoosa County citizens had voted "No" to the state's secession, once the war began, the native sons joined the ranks of the Confederate army and fought with Lee until he and his tattered troop surrendered to Grant at Appomattox on April 9, 1865.

Divided loyalties, secret societies and underground peace society movements were so widespread in Alabama that on May 8, 1864, Assistant Adjutant General H.W. Walker reported to General Braxton Bragg, "General, I returned yesterday from my tour of investigation as to the secret treasonable society alleged to exist in this State…I am satisfied that the society embraces more than half the adult males of Randolph, Coosa, and Tallapoosa Counties, a large number in Calhoun and Talladega County, and a considerable membership in some of the other counties."[47]

At the end of the war, Reconstruction dealt a heavy blow to Tallapoosa County. The county lost $125,000 on railroad bonds alone between 1868 and 1873. Cotton production dropped so severely, Tallapoosa County was ranked as one of the "strangulated counties" because of its massive debt. But better days were ahead in the mineral exploration of copper and gold mines, and Tallapoosa County had the lead in the Alabama gold belt counties.

The Lost Confederate Gold

What happened to the mints during the Civil War and to the Confederate gold?

On March 3, 1835, Congress established branch mints at New Orleans, Louisiana; Charlotte, North Carolina; and Dahlonega, Georgia. "Gold fever" was raging in Alabama, especially at Arbacoochee, where gold mining was at its peak. Thousands of miners were feverishly digging in Alabama

The Old Southwest

dirt, busting rocks and hauling ore to streams, where they sorted gold from rock. The tragic Trail of Tears lay ahead in the year 1838, followed by the discovery of gold at Hog Mountain in 1839 and at Goldville in 1842. In 1835, in the midst of skirmishes between the pioneers and the remaining Native Americans, few people could have imagined the war ahead in 1861 that would divide the North and the South, separate families, take the lives of an estimated 620,000 Americans and leave hundreds of thousands of others wounded and maimed. At the onset of fighting, the mints were closed to keep southern gold from falling into the hands of the Union army. The gold went into the treasury of the Confederate States to fund the troops and was moved to several locations to keep Union troops from finding and seizing the treasure. After the war, unsuccessful searches for the "lost" Confederate gold fueled the imagination of gold seekers. A story emerged that George Trenholm, a shipping and banking magnate, stole the gold while serving as the treasurer of the Confederacy. If the legend is true, Trenholm may have been the inspiration for the character Rhett Butler in Margaret Mitchell's book and the classic movie *Gone with the Wind*.[48] Myths and legends abound as to where the Confederate gold was hidden. Although its location remains a mystery, a number of stories emerged surrounding the loss of the gold.

In May 1861, Jefferson Davis moved the capital of the Confederacy from Montgomery, Alabama, to Richmond, Virginia, to promote support in the area. In April 1865, Davis and his supporters evacuated Richmond to avoid capture by the Union army. The Confederate officials boarded the first of two trains; the second was loaded with "special cargo," gold and silver. The trains stopped at Danville. Davis and his men traveled south on horseback. Captain Parker, who was in charge of transporting the Confederate treasure, stored it in coffee cans and sugar and flour sacks. Eventually, Brigadier General Basil Duke relieved Captain Parker and, with around one thousand men and a caravan of six wagons, headed south. Reportedly, "bushwhackers" robbed the wagons and took as much gold as they could load into saddlebags and sacks. The legend is that they buried the rest in different places in Wilkes, Georgia.

Martha Mizell Puckett's *Snow White Sands* suggests Jefferson Davis divided the gold among everyone present at the last meeting at Washington, Georgia, and left its use to them. Another account maintains Jefferson Davis placed the entire Confederate treasury into the care of Sylvester Mumford, whose family donated scholarships to the children of Brantley County with the remainder used to establish the Thornwell Orphanage in Clinton, South Carolina.[49]

Alabama Gold

During the Civil War, private mint coins made by the Bechtler family of North Carolina (1830–1852) were endorsed by the Confederate government and circulated for use. The Bechtler mint started in North Carolina after the first major discovery of gold in the antebellum South. Circulation of gold coins continued after the Legal Tender Act of 1862 outlawed private minting of coins and after the Civil War ended.[50]

Copper Excitement

Beginning as early as 1853–54, Cornish miners began to drift into Alabama from the Ducktown copper mines in Tennessee. The prospectors dug shafts one fathom square to a depth of ten fathoms, after which, if no copper was found, they abandoned the shaft. The Cornish miners prospected principally iron gossans, which were considered to be surface indications of copper. While they did much useless work looking for copper between 1859 and 1861, some of them found small amounts of gold deposits. However, they became fearful of being drafted into the Civil War, abandoned their search for copper and went back to Europe.[51]

Long before prospectors searched for copper in the southern Appalachian Mountains, Native Americans highly valued copper and used it to create religious symbols, such as the "Rogan Plate" found by John P. Rogan at the Etowah Site in Alabama in 1883.[52] Due to copper's malleable and durable qualities, Native Americans used the metal for practical and ornamental purposes. Evidence of the Creeks' use of copper in crafting tools and jewelry can be observed in the Indian artifacts uncovered in east-central Alabama. In this region, no Native American artifacts crafted from gold were found to support the persistent myth that Indians had great stores of gold they hid from early explorers and settlers.

In studies of rocks and minerals, state geologist Lewis Dean recorded the following locations of copper deposits in Alabama:

> *Copper occurs in Alabama as chalcopyrite (copper iron sulfide) and is found in association with massive sulfide deposits in Clay and Cleburne counties. Exploration for copper began in the northern Alabama Piedmont in the 1850s near the towns of Pyriton and Millerville in Clay County and the Stone Hill mine (also known as "Wood's Copper Mine") in Cleburne*

The Old Southwest

County. *Copper ore was mined at Stone Hill from 1874 to 1879 and shipped out of state for smelting.*[53]

In the search for copper in the 1880s, prospectors made a few discoveries of gold, creating a revival of interest in gold mining. Shafts were sunk in some of the well-known gold veins; machinery was brought in for equipping mills, and systematic mining was attempted, but the gold below the level of weathering was found impractical of recovery by amalgamation. The height of the gold veins and abundance of hard quartz ore at Hog Mountain attracted miners to the mountain. A boom in production occurred when cyaniding was introduced in Alabama in 1903 by the Hillabee Gold Mine Company. Hog Mountain was the only mine in the state where cyanide was used extensively. The result was a sudden increase in Alabama's total gold production in 1904.[54]

Iron and Steel

The year 1874 also was important to the Tallapoosa County residents for a reason other than the revival of interest in mining copper and gold. The town of Youngsville now was officially Alexander City, named for the president of the railroad Edward Porter Alexander. On the appointed day, a large crowd of townspeople gathered at the Alexander City Depot anticipating the arrival of the steam engine named *Captain Simon Suggs*. The name honored Johnson Jones Hooper's Old Southwest frontier character and Bird Young, who was said to have inspired the character and was a descendant of the Youngsville founding family. Most of the townspeople had never seen a train, so they came prepared to relax and enjoy the festive occasion with friends and family.

The local ladies in long dresses and hats and the men in their best attire were laden with umbrellas, tablecloths and picnic baskets, chatting with neighbors as they watched and listened for the first sound of the train. They heard in the distance the loud sound of wheels clacking against the tracks and, soon after, saw smoke bellowing out of the approaching engine. When the conductor blew the horn to warn people and animals to clear the track, the startled crowd ran in alarm, scattering chairs and picnic items behind them as they retreated to a safe distance. Despite the townspeople's initial fright, "meeting the train" became a regular event for them. In their leisure,

Alabama Gold

they gathered to observe the trains' arrival with cargo and passengers and see who might be embarking on an excursion outside the small town, now known to locals as Alex City.

The Savannah & Memphis Railroad brought opportunity to towns and communities to connect with the larger world of commerce and entertainment. Local merchants could order and receive shipments of dry goods and farming and mining equipment to sale at their stores. The railroad made it possible for the mine owners to have the latest equipment shipped to Alexander City, where they could load machines and supplies into wagons for delivery to mining sites. As mining activity increased in Tallapoosa County, companies such as the Hillabee Gold Mining Company regularly shipped ore samples to the St. Louis mint for evaluation. Potential investors, engineers and miners rode the passenger trains into Alexander City and rode buggies into rural Tallapoosa County to inspect the numerous gold mines and enjoy the beauty of the surrounding countryside.

The years during and immediately following the Civil War consumed the energy, resources and economy of the South. Gold mining activity consisted of the occasional panning by farmers to supplement their farm income. They traded their "pinches" of gold locally for farm supplies and household staples. No record exists of the amount of gold individuals retrieved from the streams and hills in the gold rush, but through the years, rumors persisted that the old mining sites were still rich with gold. Those rumors reached

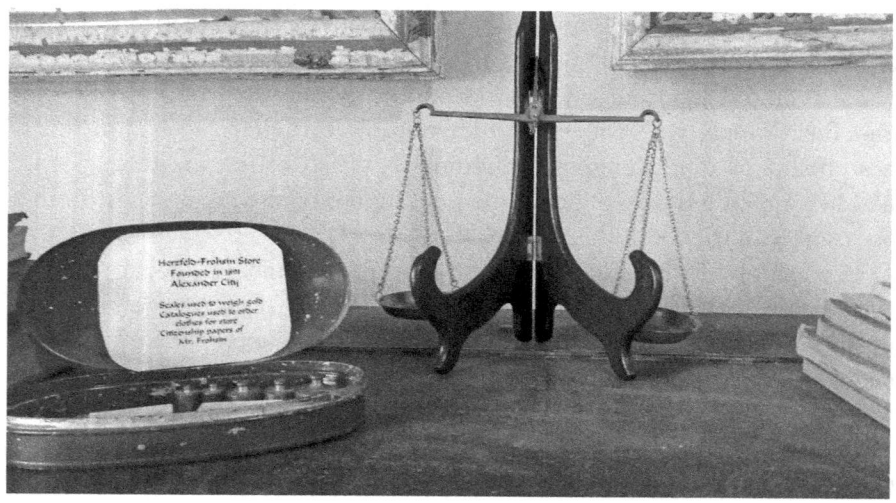

Weights and a stand used to weigh gold in the Herzfeld-Frohsin store, founded in 1891 in Alexander City. *Courtesy of Tallapoosa County Historical Museum.*

The Old Southwest

beyond the hills of Tallapoosa County and stimulated interest in Alabama gold as far away as New York and London.

During this time, the *Vidette* reported an interest expressed by potential investors in the county's mining potential.

> *Men in the financial capital of the United States were aware of the gold in Tallapoosa County. A letter to Colonel Dean from E.M. Morgan of R.A. Ammons and Company, bankers and brokers of #2 Wall Street, New York dated December 2, 1887 states "Regarding that Birdsong and Jones property. Mr. Roudebush leaves tomorrow for London and from cables and letters which we have received we are confident that the property will be taken by some people over there."*[55]

Gunfights and Gold Mining

Even in the 1880s, elements of the Old Southwest frontier remained. The White Elephant and Allen's were two saloons where men gathered to socialize and promote the local whiskey industry. Determined to establish law and order, citizens donated money to build the first courthouse in Alexander City in 1889. Fights and shootouts in the saloons made the local paper. The following account is of a dispute and gunfight between two friends and well-respected men:

> *A shooting affray occurred on Tues., 7:00 pm at Allen's Saloon between two men who were considered to be friends, Dr. H.J. Cameron in charge of mineral developments at the Romanoff Land and Mining Company and Dr. P.D. Mahoney, a well-known oculist, from Covington Georgia. The two men exchanged about nine shots. Both were taken to the Polk Hotel, and their lives were saved by local doctors. The reason for the shooting was not known.*[56]

The editor of the *Vidette* R.A. Posey reported renewed interest in gold mining in Tallapoosa County:

> *In 1888, Colonel Moore at the Hog Mountain gold mine was rapidly turning out what he called the "finest specimens of gold." He had been, he said, in the gold business all his life and made it a profession, and*

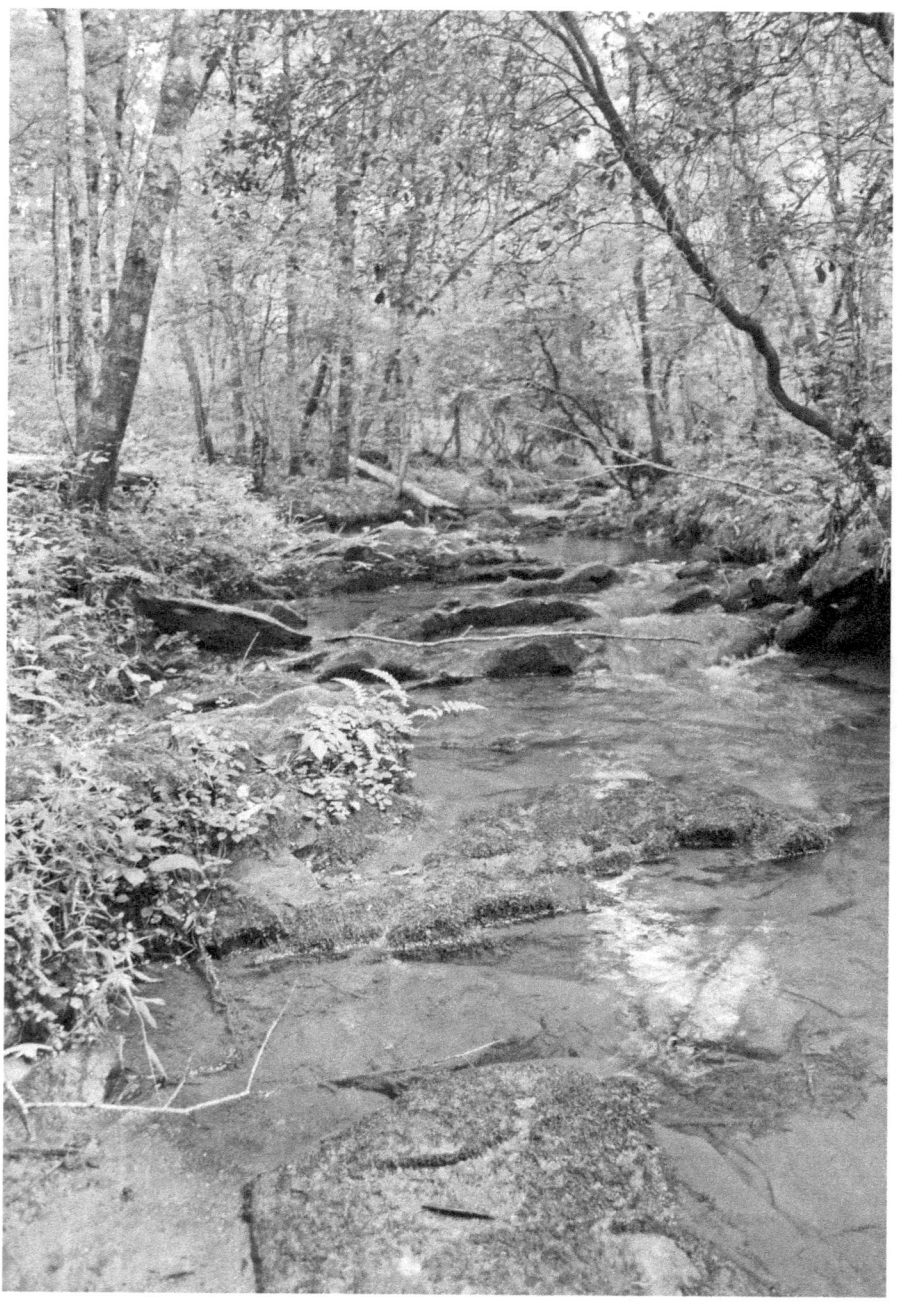

Moore's Branch was named for A.H. Moore, who mined on Hog Mountain in the 1880s before T.H. Aldrich purchased the property. *Courtesy of Peggy Jackson Walls.*

The Old Southwest

nowhere were there finer mines or ores richer with gold than those in Tallapoosa County.

There were rumors of capitalists from New York coming to invest in gold production here. The area boasted, "We have hidden beneath our hills and mountains treasures that will in the near future be developed and make this the 'Gold Country' of the South. Alexander City, nearest to the most productive mines, began calling itself the 'Gold City.'"[57]

The January 9, 1890 *Vidette* concluded, "The outlook for a building boom in the Gold City, the coming spring is very promising." In the same issue, an ad for gold mining machinery reads:

I will sell on Wednesday the 22nd day of Jan 1890 at the 10 Acre Gold Mine in Tallapoosa County, Ala., to the highest bidder for cash, a ten stamp battery, copper plates and fixtures, together with an engine and boiler. This property is known as the A.H. Moore machinery and is sold to satisfy a note for the purchase money. This December the 26th, 1886. John Cross. John A. Terrell, Atty.

A note about business improvements in Alexander City appears besides A.H. Moore's sale notice.

The Proprietors of the "White Elephant Saloon," Messers Clark & Howell have had erected this week, a substantial awning in front of their saloon. We are informed no liquor [business] that have applied for license for the year 1890, in this county has been refused. This is right, give us more of the "good stuff" and a little better quality.

Progress and Setbacks

While panning in the creeks remained a common activity for farmers and individual prospectors, larger enterprises were being planned for the Hog Mountain mine. The Alexander City newspaper forecasted a positive result from the implementation and use of new equipment and materials arriving at the train depot en route to Hog Mountain:

ALABAMA GOLD

> *The view ahead for Alexander City from the beginning of the 20th century was good. The Romanoff Land and Mining Company had entered into a contract with Mecklinburg Iron Company to erect at the Hog Mountain mines a $10,000 plant with modern machinery that it claimed would save a much larger percent of the gold found than had been possible with the older cruder methods. Several carloads of machinery and building materials came in via the railroad and were transported from Alexander City to the mines.*[58]

In 1899, electricity came to Alexander City, with around three hundred homes subscribing for power. In downtown Alexander City, a light was installed on the town square, one at the courthouse, one on Church Street and one at the Alliance Warehouse. The accessibility of power in Alexander City had another important consequence. It was now possible to run lines to the Hog Mountain mine to supply the power needed for operating the new machinery. The Alexander City Council had plans for a waterworks and better firefighting equipment when a small fire started that burned the town to the ground. By the time the first bucket of water was splashed on the fire, the wind had changed and swept the flames through downtown, burning the wooden structures: the depot, the boxcars on the tracks, the tracks, the livery stables and the animals in them, the wooden sidewalks, the Alexander City Bank, the Citizens Banks, the courthouse, the hotel, the post office, the newspaper office, stores, churches and nearby homes. Within a week, stores were set up in temporary booths much like a street fair and ready for business. People from surrounding counties and cities—even from as far away as Columbus and Atlanta—sent money to help rebuild the town. Farmers came with their teams of horses on scheduled days to help clean up debris. The town was rebuilt in about four months with a renewed vision for its future as a thriving, prosperous city.

Alexander City became famous, not as "gold city," but as a thriving mill town, home first to the Alexander City Mill. In 1902, Russell Manufacturing Company was established, followed in 1919 by Avondale Mills, which bought the old Alexander City Mill. For over a century, Russell and Avondale Mills provided thousands of jobs to people in Alexander City and in outlying communities. Business was booming on both sides of the railroad track after the town was rebuilt. People were encouraged to spend their money locally to support the town, its schools and churches. The railroad and businesses in Alexander City were essential to the gold mining industry in Tallapoosa County and the mines' success, in turn, strengthened the town. Local

The Old Southwest

newspapers promoted the gold mines with updates of any activity within the county. The following report in 1904 read:

> *The ever present gold mining activity was back in the news with a report by Dr. H.J. Cameron, president and general manager of the Hillabee Gold Mining Company, which had been operating a 10 stamp mill and mining gold ore in paying quantities for the past two years. Dr. Cameron, a mining engineer, who lived in Alexander City and had worked in Mexico, France and other foreign countries, said that there were thousands of acres of land in Tallapoosa, Clay and Cleburne counties that could be mined with paying results. A party of the directors of Schloss Coal and Iron Company of Birmingham came into town by special train in the early summer of 1904 to spend the day in the gold field of Tallapoosa County. There was activity at the Ely pit near the Hog Mountain croppings, and nearby openings made by T.H. Aldrich were promising. It was said that from these openings there "will be gold ore in such riches as to turn the eyes of the whole industrial section this way." And there arrived by freight the heaviest load of machinery ever carried out of Alexander City. Will Jarvis of Cow Pens and three other drivers of a six-yoke team hauled the heavy machinery, which was a big tube mill weighing ten tons for the Hog Mountain gold mine, the 15 miles to the mine. Over $700 in one week was paid for freight bringing mining machinery to Alexander city for delivery to the mines. The Birmingham News reported the total output of gold in 1905 in Alabama was $45,500, an increase of $16,500 over the previous year, and nearly all of this amount was mined by T.H. Aldrich, Sr. and Jr. from their property in Tallapoosa County, fifteen miles north of Alexander City.*[59]

The scarcity of local labor created a problem in the mining operations and resulted in hiring foreign workers. In March 1906, a party of six Italians arrived in Alexander City by train and were conveyed to the gold mines at Cow Pens. These were perhaps the first Italian laborers ever brought to this section, but it was expected there would be many others.

The Dutch Bend mines were a popular place for many outings. According to the townspeople, there was a no more picturesque spot in Alabama, and it was fashionable to entertain guests with a picnic or overnight camping. Camping trips were a regular event:

> *One July, when the weather was perfect, a jolly party in honor of visiting Vollie Askew and her sister of West Point spent two days camping at Dutch*

ALABAMA GOLD

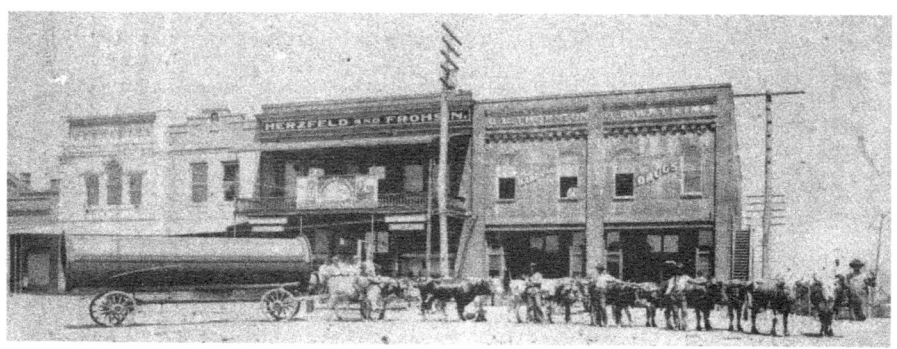

This ten-ton tube equipment headed to Hog Mountain was drawn by oxen to use at the gold mine. *Courtesy of Peggy Jackson Walls.*

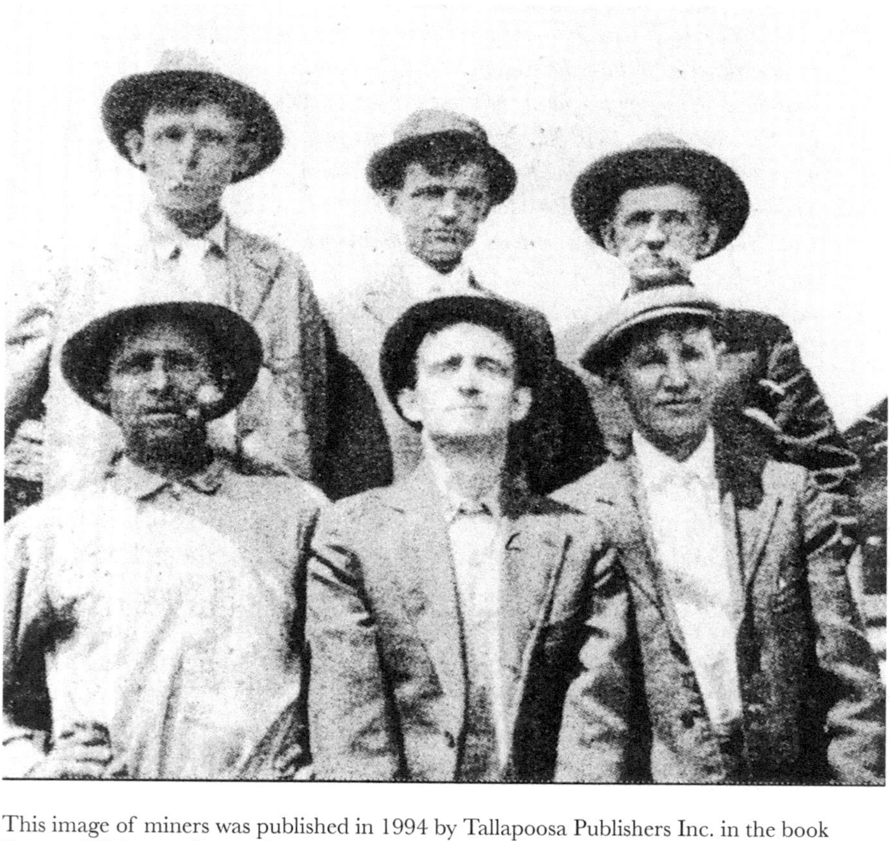

This image of miners was published in 1994 by Tallapoosa Publishers Inc. in the book *Images of Tallapoosa County. Courtesy of Kenneth Boone, Tallapoosa Publishers, Inc.*

The Old Southwest

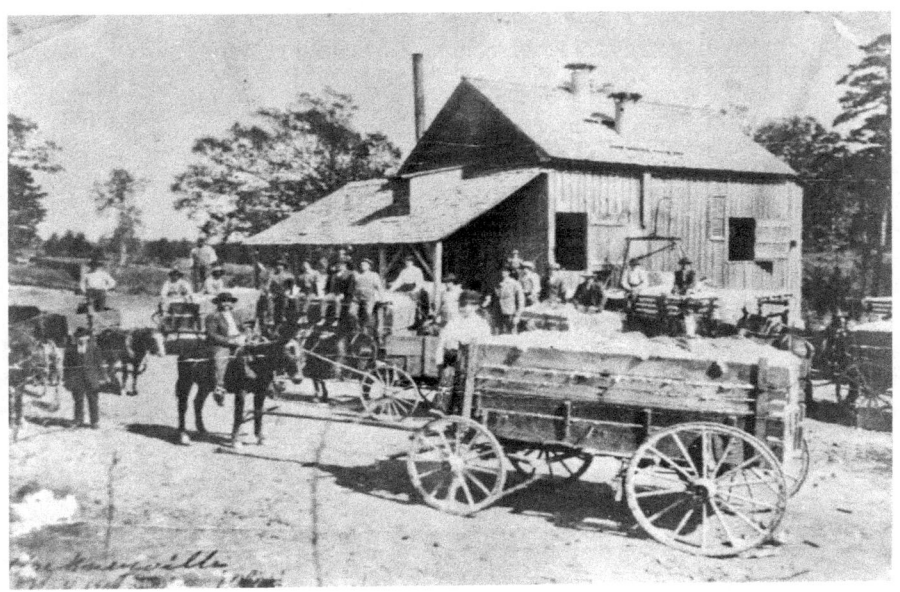

Farmers bring their cotton crops to the gin in downtown Hackneyville, 1910. *Courtesy of David and Karen Daniel.*

Bend. The group included L.B. Coley, W.L. Radney, T.S. Christian, Jr., and Charles Dean. On another occasion, busy Dr. T.H. Street went up to Dutch Bend with his wife and some guests for two weeks of rest and relaxation.[60]

Dr. Street was the doctor for the Hillabee Gold Mining Company, but his younger associate Dr. James Cameron made many trips to the Hog Mountain mine and surrounding communities to take care of the miners and their families.

With gold mining activity at their backdoor, most farmers depended on cotton for a regular income. There was a gin in downtown Hackneyville in 1910 where farmers brought their cotton to be processed. Farmers took their cotton yield to Alexander City and to the nearest farm markets. Built in 1840, the Central Plank Road started in Montgomery and passed through Wetumpka into Tallapoosa County. The old plank road was traveled regularly by farmers taking their goods to the marketplace in Wetumpka.

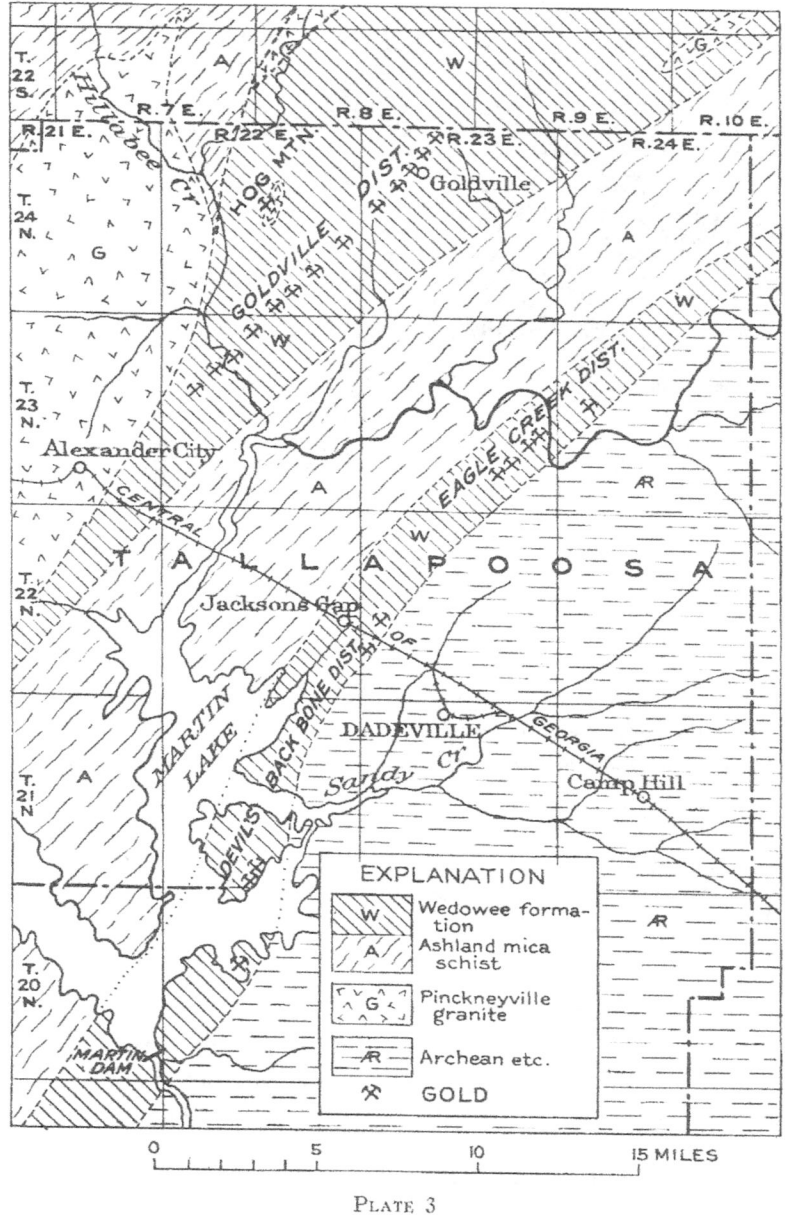

PLATE 3

Four gold mining districts in Tallapoosa County were the Devil's Backbone, Eagle Creek, Goldville and Hog Mountain. *Courtesy of John F. Farrow.*

5
Tallapoosa County: "Gold Country"

THE DEVIL'S BACKBONE DISTRICT

The Devil's Backbone District in the Weodowee formation extends as far west of the Tallapoosa River as Elmore County. East of the river, the belt continues in a northeast direction into Chambers County. Rocks in the belt are primarily slate, phyllite, quartzite and schist of sedimentary origin. In the middle section, the beds dip at steep angles toward the southeast with heavy sections of quartzites exposed in a ridge known as the Devil's Backbone. Antebellum miners worked in the auriferous gravels along the streams, without gaining any large amounts of gold and now are covered by Lake Martin. In 1845, state geologist Michael Tuomey inspected the Silver Hill Gold Mine, located on Copper Creek. He reported a yield of ninety-six dollars per ton of ore, noting the main vein was about two feet thick on the surface and richest at twelve feet. Tuomey observed equipment of six stamps and a badly constructed Burke Rocker.

In 1891, state geologist William B. Phillips inspected the Silver Hill mine and reported the mine was currently being operated by Major Parmalee, who later moved the mining equipment to the Gregory Hill Gold mine and operated a fifteen-stamp mill there. Parmalee reported removing $80,000 in gold value out of Silver Hill, Dent Hill and Gregory Hill at a cost of $100,000.[61] Major Parmalee operated mines in the Devil's Backbone District for a decade or more from the early 1880s through the 1890s.

Alabama Gold

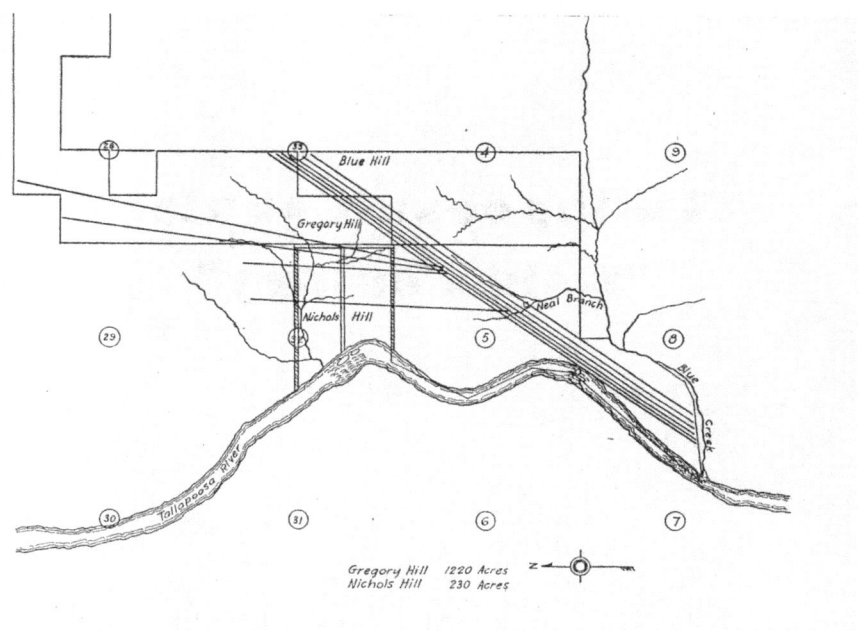

The sketch outlines locations of Blue Hill, Gregory and Silver Hill Mines in southern Tallapoosa County. *Courtesy of John F. Farrow.*

From the sparse sources available from this period, Phillips's reports of the scattered mining activity in Tallapoosa County provide insight into mining methods, equipment and the ingenuity of the mine owners and engineers. Almost no resources exist beyond the geological surveys and engineering reports created primarily for other geologists and engineers. Although they were not created for general audiences or casual readers, Phillips's inspections and subsequent surveys record the gold mining activity taking place during the last two decades of the nineteen century.

In 1882, Major C.H. Parmalee from New York used various methods to mine ore at the Gregory Hill mine, first by tunneling and later by quarrying, due to the difficulty of separating slate and quartz in the tunneling method. In the 1880s, Major Parmalee also operated a fifteen-stamp mill on the south at Black Branch, north of Gregory Hill. Bill Phillips recalled witnessing the operation in the 1890s, when he was twelve years old. Phillips described the transportation of ore in two wagon-sized carts from Gregory Hill to the mill by gravity. "Ropes interconnected the front to rear

The Old Southwest

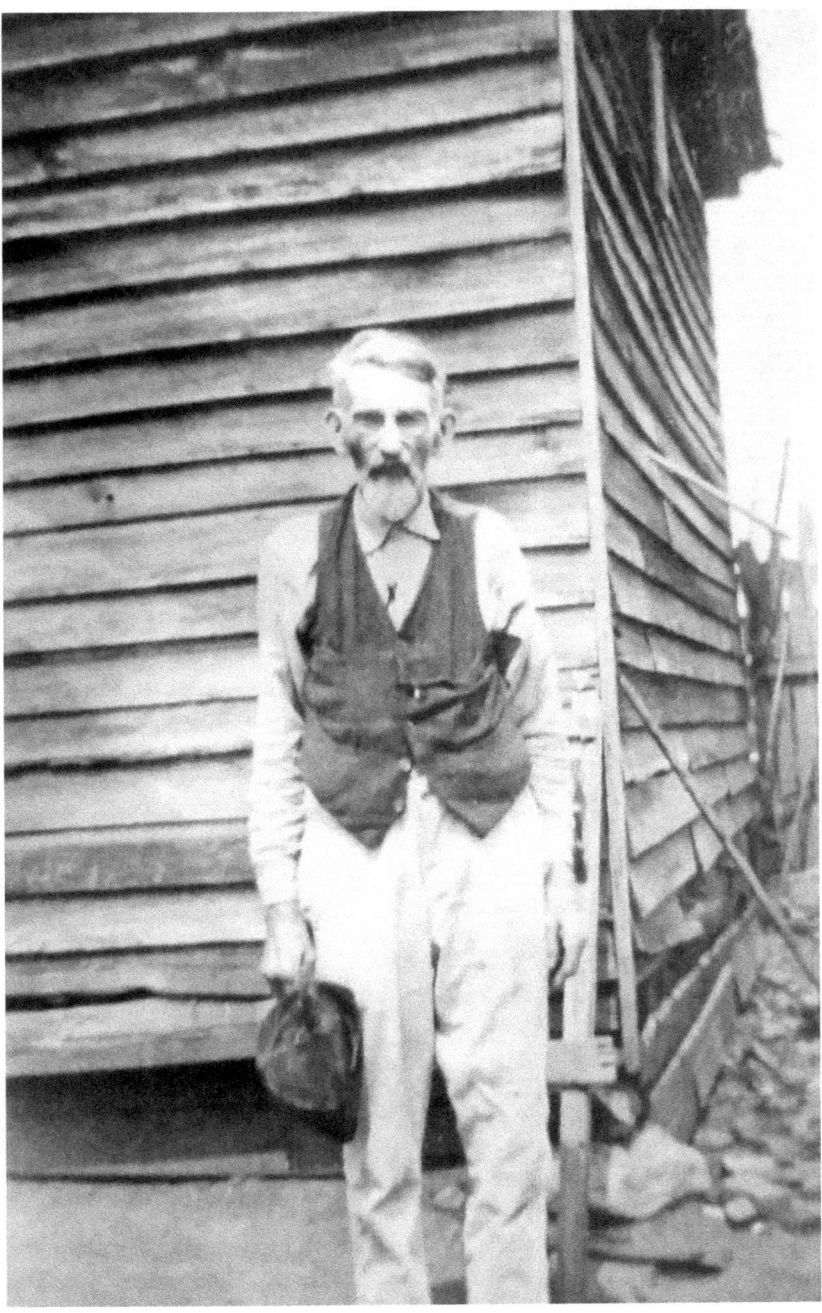

Thomas Tyler "Tom T." Farrow, gold miner and owner of the Farrow gold mines. *Courtesy of Tallapoosee County Historical Museum.*

of each cart to form a loop so that when one cart was loaded and the rider loosened the brake, it would go downhill to the mill and simultaneously pull the empty cart up the hill to be reloaded."[62]

Judge C.J. Coley, probate judge, banker and humanitarian, provided another perspective to gold mining in Tallapoosa County through his friendship with the Farrow family and research as a trustee of the Alabama Department of Archives and History in Montgomery. Coley shared his research in articles and lectures to historical organizations, such as the Alabama Historical Association.

The Farrow mines in the Devil's Backbone District were well known, having been in operation from the 1880s through the early 1920s. Thomas Taylor Farrow established the Farrow Gold Mining Company in the Suzanna community. He operated a five-stamp mill, where large metal plates, powered by water, crushed the rock taken from the Farrow mine. In April, 1894, The Engineering and Mining Journal published a report on gold mining in Tallapoosa County, describing the stamping

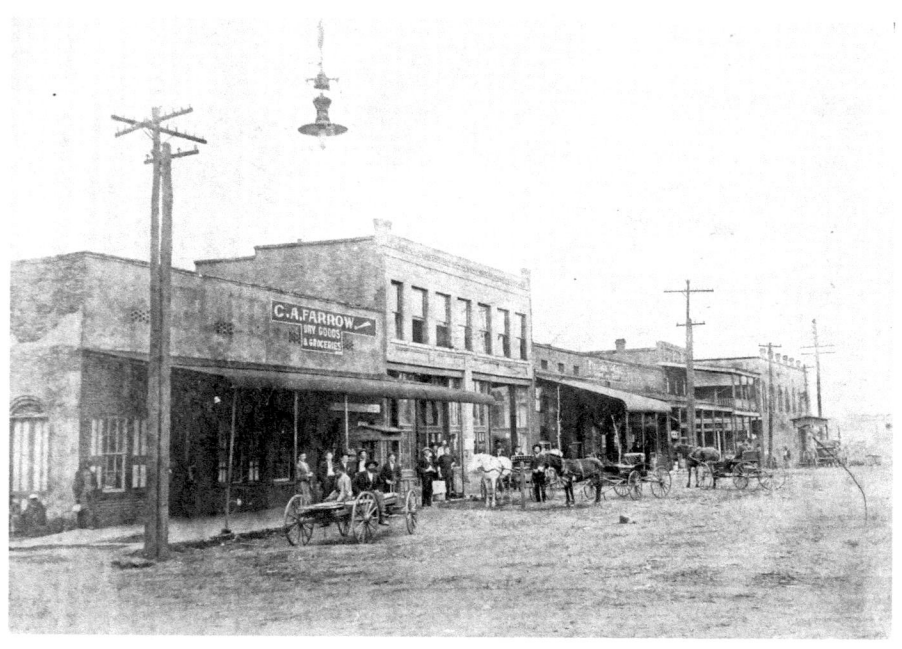

"C.A. Farrow Dry Goods & Groceries" was in operation when horses and buggies were the main mode of transportation. *Courtesy of Tallapoosee County Historical Museum.*

The Old Southwest

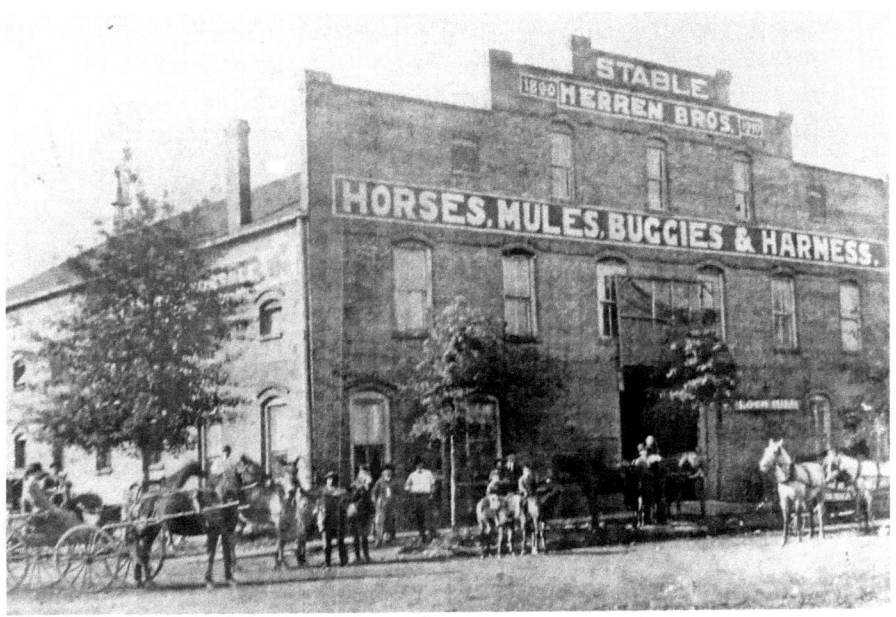

Herren Brothers business, established 1890, was still in operation when the picture was taken in 1910. *Courtesy of Tallapoosee County Historical Museum.*

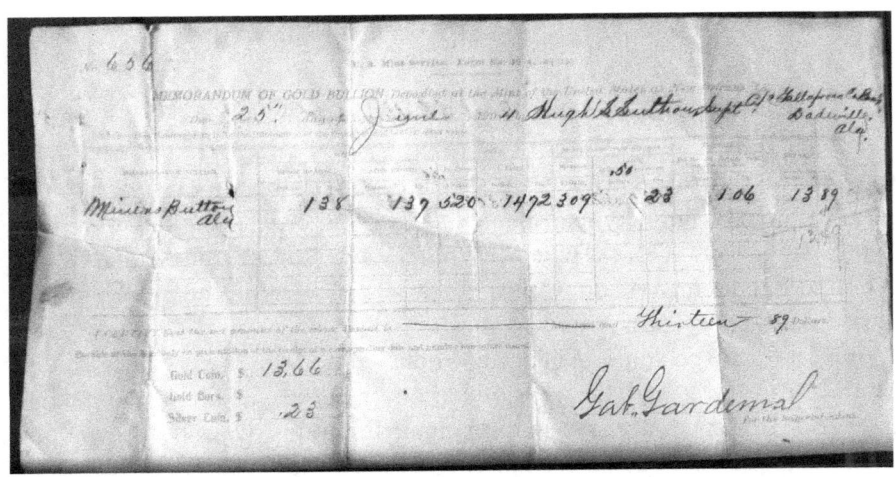

A memorandum of gold bullion deposited on June 25, 1904. *US Mint Service Form. no. 42; courtesy of John F. Farrow.*

process, "At Farrow Hill, the owner Mr. Thomas T. Farrow, has recently put in iron stamps to replace the ten wooden stamps he has been using in the past. The ore body on his property is not as highly graphic as at other mines in the area and bears more the characteristics of free milling...of the ore in the Southern states."[63]

Located ten miles south of the county seat of Dadeville, the Farrow Gold Mining Company operated for 25 years following the War Between the States. Thomas Taylor Farrow built his own mine machinery, including a self-powered railroad system. Gold produced from the operations was minted in New Orleans, but to earn a dollar profit required spending three.[64]

EAGLE CREEK

Although gold mining in the Eagle Creek District was not as extensive as mining in southern Tallapoosa County or in the northeast districts of Goldville and Hog Mountain, numerous operations took place dating back to the 1840s. Geologist William Brewer reported finding tunnels and evidence of numerous mining operations on properties, such as the Hammock Mill, the Jennings property and the Greer property.

NEAL BRANCH MINING COMPANY

One of the last mines to operate in southern Tallapoosa County was the Neal Branch Mining Company, known as a "Mull Mill," producing twenty-five to thirty tons of ore a day. Mr. Sharp was manager and Mr. Goss was the superintendent. William Coley Farrow also worked in the mine before it closed due to operation costs.

In 1892, state geologist William B. Phillips reported, "Two samples of 25 pounds from Gregory Hill were mixed and assayed yielding 0.3 ounce of gold and 0.1 ounce silver per ton." A small amount of work had been done at Blue Hill by 1892. The Blue Hill sample showed an essay of 0.4 ounce of gold and 0.13 ounce of silver per ton. Significant mining activity took place in 1915 on the Ben Kidd property under the name of Gregory Hill Gold Mining Company beside the Neal Branch.

The Old Southwest

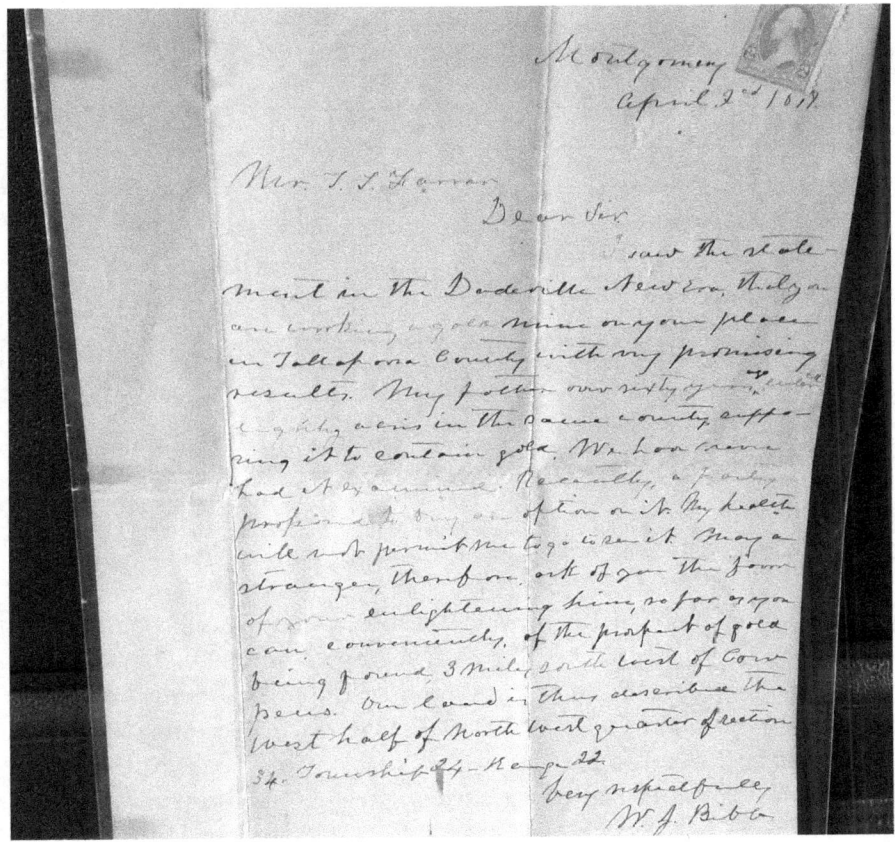

A letter to Thomas T. Farrow from 1879 in which the writer inquires about the potential of gold mines in southern Tallapoosa County. *Courtesy of John F. Farrow.*

Although much of the history of the mining in the Devil's Backbone District is anecdotal, taken from interviews with family and community members, interesting, credible articles have appeared in professional journals such as the *Engineering and Mining Journal*, 1894. Helen Pate Ross collected stories from family members and organized them into a book. John F. Farrow was one of the people she interviewed. He shared information about his family's involvement in the gold mining, and documents such as receipts, letters or reports, naming the dates mines were in operation and the approximate amount of gold retrieved from the operations. Banks were ready to finance the new companies and invest in mining operations before World War I. The Gregory Hill Gold Mining Company was formed in 1915 with a J.J. Heard serving as president and Dr. Ben E. Kidd as vice

Alabama Gold

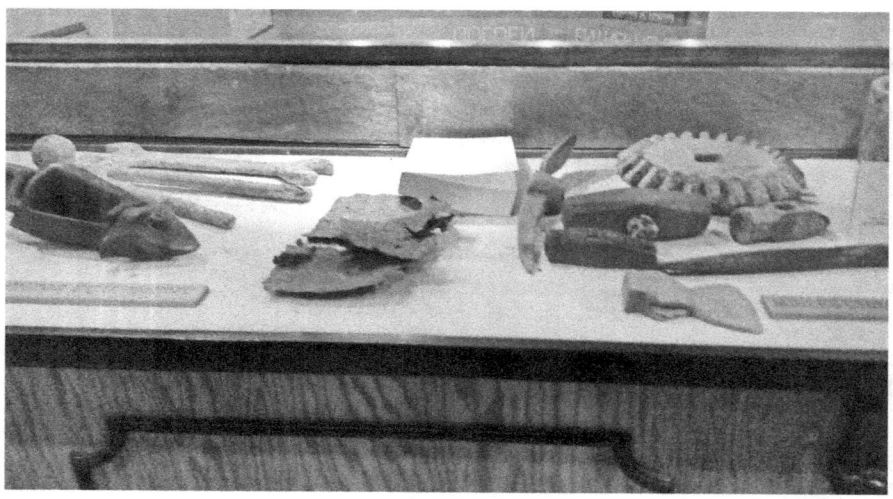

This picture of the mining tools is used with permission of Pine Mountain Gold Museum, located in Villa Rica, Georgia. *Courtesy of Peggy Jackson Walls.*

president, who deeded 1,180 acres in Tallapoosa County to the company in return for 3,000 shares ($300,000 par value) capital stock; 5,000 shares were the total authorized. A $35,000 loan obtained from Heard National Bank financed the operation. Only $10,780 of the loan went into camp development and machinery. The rest went to salaries, maintenance, legal expenses, and bond payments. An L-shaped mill building, well, cookhouse, bunkhouse and mule pen were constructed. After only nine months of operation and a total expenditure of $17,191.56, the mine closed due to World War I, which "caused the total suspension of gold mining by cutting off the supplies of mercury and cyanide, increasing wages and calling labor to other pursuits."[65]

A smaller operation was conducted on the Nichols Mine Property, located on the south side of the Tallapoosa River. The property was owned by B.E. Kidd Jr. and other heirs to the estate of B.E. Kidd. Professor William M. Brewer, assistant state geologist, recorded in Bulletin No. 5, Alabama geological survey, his evaluation and recommendations regarding the Devil's Backbone gold mines.

> *On the west side of the Tallapoosa River in Tallapoosa County there were operations at Blue Hill, Gregory Hill and Farrow Gold Mining Company, during the 1880s and 1890s.*
>
> *In traveling to the north-east of Blue Hill, I found that the gold bearing quartz, and graphitic slates were not as closely associated as in the case*

THE OLD SOUTHWEST

at Blue Hill and Gregory Hill (both in the same land section), where the two became a conglomerated mass, so intermixed and irregular in structure that it is impossible to separate the one from the other and the entire sides are quarried down, and send to the mill. This product from the mines was yielding at the time of my visit about $2.00 per ton, or, rather that amount was saved. I am of the opinion though, after panning several average samples, that at least double this amount would be saved, were some process devised for eliminating the graphite before amalgamation was attempted.[66]

OPERATION IN THE EARLY 1900s

In 1902, Ben E. Kidd Sr. purchased extensive land holdings abutting the original eighty-acre Blue Hill property. The Blue Hill Gold Mining Company leased and developed the Blue Hill mine during 1902–07. Gold mining was dormant from about 1907 until 1934, when the property was leased to Frye-Rhea Development, financed by investors in Cleveland and New York City. The Frye-Rhea enterprise was the last attempt to mine at Blue Hill. Again expenses outweighed profit, causing owners to close the mining operation.

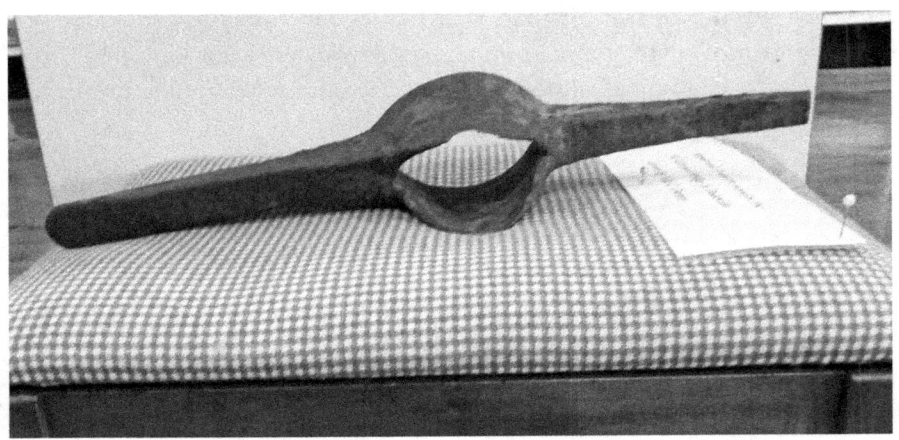

Tom T. Farrow's gold pick. He used it to prospect for gold throughout his lifetime. *Courtesy of Tallapoosee County Historical Museum.*

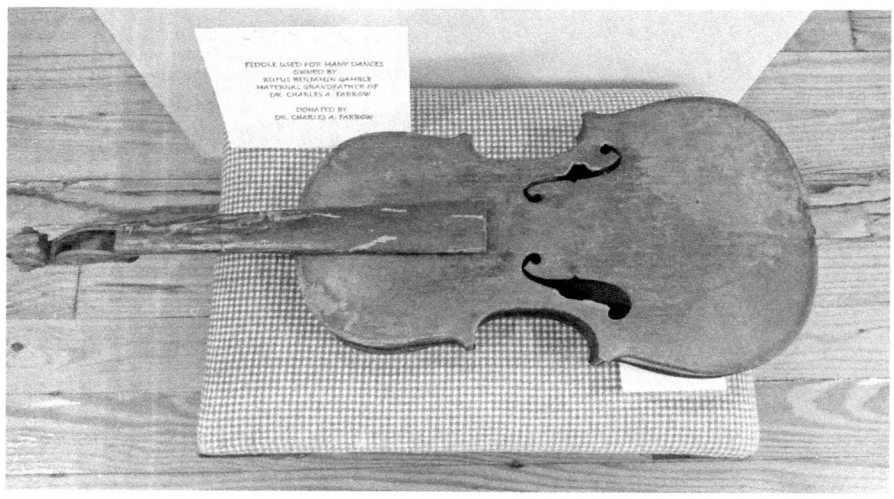

Rufus Benjamin Gamble, maternal grandparent of Dr. Charles A. Farrow, played his fiddle at local social gatherings of friends and family. *Courtesy of Tallapoosee County Historical Museum.*

Fanny Goldmine Hill

One of the stories associated with this district comes from the early years when Indians still owned some of the land in Tallapoosa County. The widow of an Indian was told she had to give her land an English name before she could register it properly. She had heard the name Fanny from settlers and liked the idea of owning a gold mine, so she registered her land under the name "Fanny Goldmine" although the land was not gold mining property.

6
Hillabee Gold Mining Company (1890–1916)

THE FAMOUS HOG MOUNTAIN

In east-central Alabama, almost the entire county of Tallapoosa is in the Weodowee Schist. But the largest occurrence of gold veins are in the tons of quartz at Hog Mountain, which is described as "the most extensively

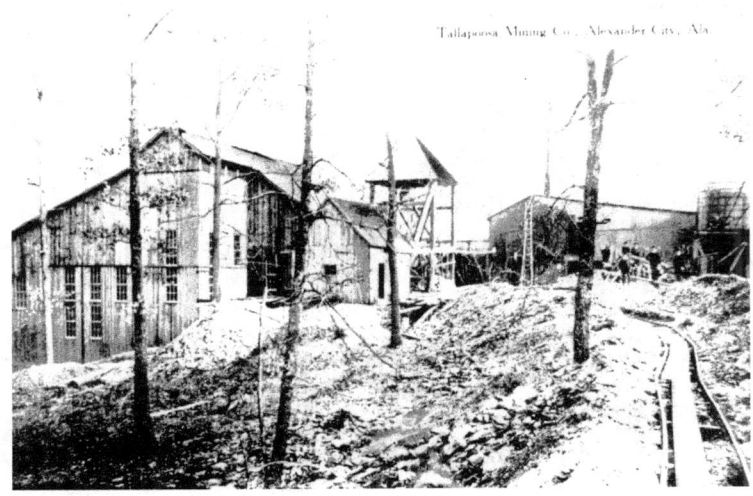

Tallapoosa Mining Company at Hog Mountain in 1911. The original image comes from the old Alexander City State Junior College Alabama Room. *Courtesy of Peggy Jackson Walls.*

Alabama Gold

developed gold mine in Alabama and is credited with a total production of about 24,000 troy ounces of gold."[67]

Three successful operations have been conducted at Hog Mountain, the first from 1839 to 1849, the second from 1893 to 1916 and the third from 1933 to 1937. The state geological surveys are the principal records of operations.

The mine was operated intermittently prior to 1890, when it passed into the ownership of T.H. Aldrich Sr. From 1893 to 1914, production totaled about $250,000. The largest part of this production came from the use of cyanide processing. The chart below shows the production of gold beginning in 1830, the year gold was discovered in Alabama and ending in 1879. However, a great deal of the gold mined in Alabama was never shipped to the mints but was instead spent locally for household, farming and mining needs. When cyaniding was introduced in 1904, the figures jumped. For additional information, see the following chart.

Value of Gold Produced in Alabama

Year	Gold Valued in Dollars
1830–79	$365,300
1880	1,000
1881	1,000
1882	3,500
1883	6,000
1884	5,000
1885	6,000
1886	4,000
1887	2,000
1888	5,600
1889	2,639
1890	2,170
1891	2,245
1892	2,419

The Old Southwest

1893	6,369
1894	4,092
1895	4,635
1896	6,495
1897	8,455
1898	6,578
1899	4,766
1900	2,618
1901	3,773
1902	2,938
1903	4,894
1904	29,288
1905	41,530
1906	24,921
1907	25,982
1908	41,208
1909	29,239
1910	33,533
1911	18,916
1912	16,724
1913	11,094
1914	11,970
1915	5,243
1916	8,650[68]

   A great deal of interest in mining the old gold mine sites resulted in increased activity and speculation that mining would be profitable. William B. Phillips examined the Hog Mountain property to determine its condition and the viability of conducting a new mining operation. As a state geologist, Phillips regularly visited the mining sites that were operating to record the activity and determine the potential of the mine. Following a visit to Hog Mountain in 1891, he recorded the following observations:

> There was a ten-stamp mill California pattern with engine and boiler on the property....He estimated that 500 tons of ore had been mined and milled....Assays from samples which Phillips took at random from the old dump gave $58.67, $6.20 and 22.73, and one from the bluish quartz with a little pyrite, taken from the Blue Vein gave $10.53... One assay gave $8.26 in gold and 36 cents in silver per ton. A barrel or ore sent to the St. Louis Works for treatment gave $39.27 in gold and 10 cents in silver per ton. Many assays made in 1886 and 1887 gave from $2.00 to $31.00 and averaged $7.50 per ton, and this was spoken of as representing the true character of the ore better than any other investigation which had been made.[69]

As president and general manager of the Hillabee Gold Mining Company, T.H. Aldrich Jr is a credible source for data about gold mining at Hog Mountain in the pre–World War I operation. Aldrich's firsthand observations about the methods and process of gold mining at Hog Mountain provide the perspective of an engineer, manager and mine

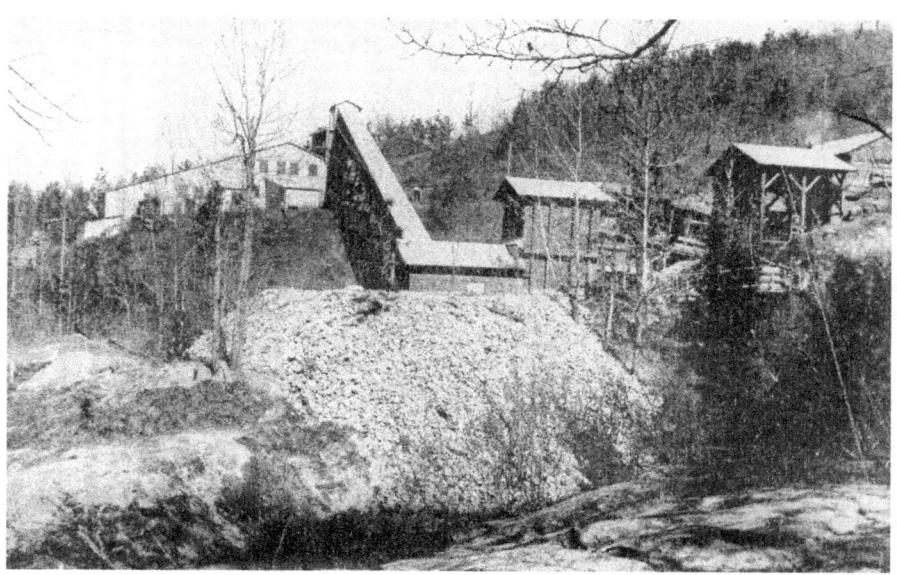

The 1935 Hog Mountain Gold Mining and Milling Company, reprinted from *The International Atlas of the World*, War History Collection, Chicago, 1946. *Courtesy of Peggy Jackson Walls.*

owner. Since water is essential to the gold mining process, the proximity of a reliable water source and equipment for transporting the water to the gold mine were primary concerns of investors and mine owners. The value of gold ore also had to justify the expense of setting up equipment and structures needed for the operation of a mine. T.H. Aldrich Jr. made the following observations about the Hog Mountain mine operation in the last decade of the nineteenth century, noting the availability of water and the value of the gold ore:

> *A 200 H. P. hydroelectric plant was built on Enitacopko [sic] Creek in 1893 to supply water to the Hog Mountain mine. Air drills and a Blake crusher were used to dislodge and crush ore. Experiments were conducted in which ore was heated in a revolving kiln. Work was done on the Blue Vein and on the Barren Vein at a depth of 110 feet and with a 15-foot sump. The value of oxidized ore was $5.00 per ton in gold and the sulphide from $6.50 to 9.00 a ton.*[70]

The evaluations of the ore and possible profits for mining encouraged T.H. Aldrich Sr. to begin preparations for a new mining operation. First, he transferred his and wife Ana's ownership of the Hog Mountain property into the Hillabee Gold Mining Company, of which T.H. Aldrich Jr. was president; J.A.P. Kennedy, vice-president; and E.J. Smyrn, an investor. Aldrich Jr. applied for a business certification and received approval from John H. Merrill, secretary of state of Alabama, on January 14, 1905. The key objective was stated: "To mine for gold and acquire construct, improve or maintain smelting furnaces, machinery and apparatus of any and all kinds necessary or deemed necessary in and about the mining for gold." Excerpts from the prospectus for the mining company follow.

> *The total paid up capital stock of said corporation is $22,000, divided in the following manner:*
>
> *T.H. Aldrich Jr.*   *$12500.00*
> *J.A.P. Kennedy*   *$12400.00*
> *E.J. Smyrn*   *$100.00*
>
> *Hillabee Gold Mine prospectus was created to provide critical information to the potential investor in regard to cost of shares, location of mine, projected costs, and profits.*

Alabama Gold

Prospectus
The Hillabee Mining Company had a capital stock of $2,000,000.00 and divided into 2,000,000 shares of the par value of $1.00 each.

Location
These lands are in Tallapoosa County, Alabama, situated about 5 miles from Alexander City, the nearest town and railroad point, on the Central of Georgia Railroad point, and reached by good wagon road.

Climate
Reference to a map of the United States shows this district on the 33 parallel north latitude, or half way [sic] between the Ohio River and the Gulf of Mexico. This insures [sic] fine climatic conditions.

LABOR: *working 11 and 12 hours per day the following is the standard wage here:*

Miners per day	$1.00 to $1.25
Laborers in mine	.50 to 1.00
Teamsters	.50 to .75
Blacksmiths	1.25 to 1.50
Helpers	.75
Carpenters	1.00 to 1.50
Engineer	1.00 to 1.25
Fireman	.75 to 1.00
All boarding themselves	

Estimated Earnings
Estimated earnings working 100 stamps. Practical experience in milling this ore has shown an average value saved of never less than $6.00 to $7.00 per ton, but to make a very conservative estimate and place the average of all the veins at $5.00 per ton would give:

100 stamps crushing 300 tons per day value $5.00 per ton	$1500.00
Cost of mining and milling 300 tons at $1.50 per ton	$450.00
Profit net for 24 hours	$1050.00
One month 26 days	$27300.00
One year, 312 days	$327600.00

Titles
The title to each and every parcel of this tract has been carefully examined by attorneys and by an Abstract Guarantee Company and the abstract

The Old Southwest

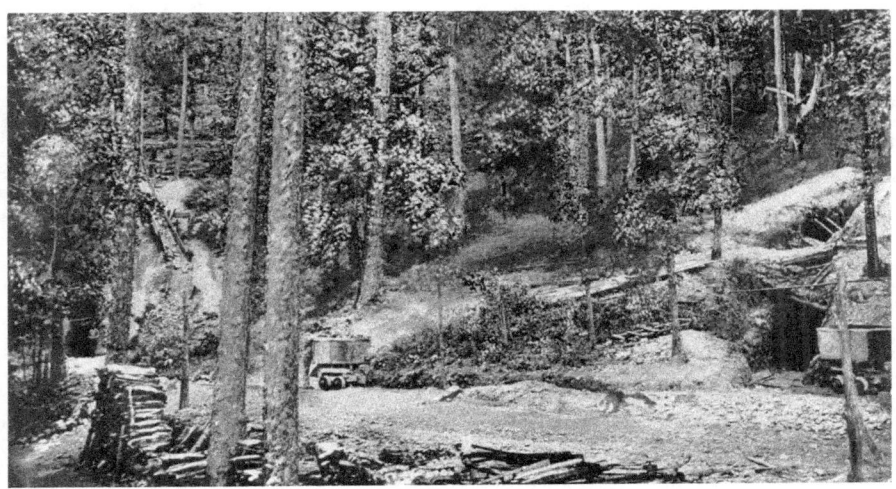

Hillabee Gold Mine at Hog Mountain, featuring the Barren Vein and Tunnel Vein, in the early 1900s. *Courtesy of David and Karen Daniel.*

this prepared and furnished to this company shows a chain of title in "fee simple" and without flaws to the grantor of this Corporation.

Equipment
At present there is a modern 20 stamp mill and chlorinating plant steam power...200 feet from Hillabee River, 5 buildings for the officers in charge at the mines, office and laboratory, all of which are well finished and furnished, also five buildings or dormitories for employees, barns and carriage houses, black smith and wood working shops, complete line of tools and all purposes required in mines, boarding houses and commissary building, two large pumps, 16,000 and 10,000 gallon tanks above mills. Other tanks for special purposes. Buildings are furnished with water by pipe lines hoisting machinery, mining cars, telephone, etc.[71]

Hillabee Gold Mining Company Closes

The mine closed in 1916 due to shortages in chemicals and labor, which was particularly drawn away into military service or employment at greatly increased wages. No other reasons were given for the mine's closing. The quantity and quality of ore remained high.

Alabama Gold

After closing the Hillabee Gold Mine, T.H. Aldrich moved to Pine Mountain, Georgia, where he leased the mining rights to Stockmar gold mine from the Stockmar family. He moved his complete cyanidation plant and equipment to Pine Mountain area. In January 1915, he started successful operations and continued so through December 1917.

> *During the last few months of 1914, Aldrich installed the cyanidation plant complete, as was transported from the Hog Mountain mine. The two leaching tanks were of Cypress wood staves and capacity of 150 tons each. The plant was erected about half way [sic] down the North Slope of the Hill and ore from mine was brought direct from tunnels and drifts onto the tanks by steel rail and dump cars of 1-ton capacity each.*

Part II

SURVIVING THE DEPRESSION: "GRINDING STONE INTO BREAD"

Ode to the Miners

*They walked the distance from clapboard and timber houses
past furrowed fields dotted with white cotton tops like
patches of small clouds cluttering the blue sky,
past morning glories, dew damp,
trailing rail fences round the neighbors' pastureland.*

*Father, son, brother in practiced step traverse the dirt road,
a unified front against the hunger and denials of the '30s.
They walked the distance to the mountaintop to excavate tons
of rock from the earth's depth, pulverize its hardness,
and make stone into bread.*

—Peggy Jackson Walls

7

The Hog Mountain Mining and Milling Company

1933-1937

News of the possible reopening of the Hog Mountain gold mine came to the rural farming communities at the best possible time—during the Great Depression. By 1929, the boom of the cotton industry had ended, with prices plunging to the lowest levels of the 1800s. The cotton market, once described as "Alabama's Gold," now was saturated with surplus goods and few buyers. The size of cotton fields shrank as the profit from the sale of cotton fell. Alabama's unemployment rate rose to 25 percent, with no significant improvement in the 1930s.[72] Few jobs were available in the cities and even fewer in rural areas. The peckerwood sawmills that moved from place to place offered a limited number of jobs and paid workers only fifty to seventy-five cents a day. The owner-operated farms were the mainstay of the rural community's economy. Families were large, and everyone was expected to do his or her part in the cycle of planting, growing and harvesting of food that would be cooked and served immediately or preserved for the winter months. Large fields of peas, squash, okra, tomatoes, melons, corn, millet and cane kept everyone busy in planting and harvesting time. Farmers raised hay to feed the cows, mules and horses. During harvesting season, neighbors worked together when families had emergencies and needed extra help.

Sharecrop farming was common. The sharecropper families were provided with a house to live in while they worked the farmer's fields. The houses were little more than one- or two-bedroom shacks, constructed with pine lumber from nearby forests. Through cracks in the floors, the farmer and family members could observe chickens pecking for worms or seeds, where dogs and cats took shelter from the rain or the summer heat. Fireplaces with stone hearths provided small relief from the winter cold as much heat escaped through the chimney openings and through slits in the pine slab walls.

But there was comfort in knowing that no matter how cold the weather might be, at least Alabama winters were short; with spring came milder weather and the start of a new farming cycle. The sharecroppers charged staples throughout the year at the country store with the understanding they would pay their debt when crops were harvested and sold. In the fall, when they were paid, they made their way to the country store, paid off the year's debt and perhaps experienced a temporary optimism that their life could change for the better. But few sharecropper families, black or white, managed to escape the lien system. They moved into houses that belonged to the farmer, worked in his fields and relied on a share of the profit from this enterprise. In this manner, they reared families and lived their lives in the hope that the crops would do well and that there would be buyers to purchase the fruits of their labor. The farm owner lived by the same cycle, one or two seasons of bad crops away from the same poverty. As bleak as the chances were for a different and better future, sharecrop farming provided a means of surviving.

To feed his family, the farmer relied on farming, hunting, fishing and slaughter of farm animals for pork and beef. Essential farm animals included chickens for eggs, a cow for milk and a mule or two for plowing fields. In the winter, wood was cut and hauled for use in the fireplace, wood heater and stove range. Nothing was wasted. Flour and feed sack cloth was used to sew clothing, sheets and pillowcases. Such were the economic circumstances in the small, rural farm communities during the Depression.

After President Roosevelt lifted the embargo on foreign shipments of gold, a 50 percent jump occurred in the value of gold with prices rising from twenty to thirty dollars an ounce. Investors and mine owners began inspecting the old mining sites and evaluating the potential profit. Due to the Hog Mountain Mine's large production of gold prior to World War I, its reopening seemed "inevitable."

When news reached the mining communities in northeast Tallapoosa County that the old Hog Mountain gold mine would reopen, residents eagerly anticipated the prospect of regular-paying jobs. The *Alexander City Outlook* published the announcement by P.S. Gardner, president of the Hog Mountain Mining and Milling Company: "Gold Mine Here to Open by Christmas to employ about 75."

Surviving the Depression

The *Alexander City Outlook* from August 3, 1933: "Reopening of Old Hog Mountain Gold Mine." *Courtesy of Tallapoosa County Probate Records, Dadeville, Alabama.*

The Hog Mountain Mine will be in operation by Christmas according to announcement made by P.S. Gardner, Jr. son of P.S. Gardner, president of the Hog Mountain Mining and refining Co. The gold mine will employ 75 to 80 men. It is situated 14 and one half miles east of Alexander City and about six miles from Goldville.

Alabama Gold

Preparations for beginning of mining operations are being carried forward as fast as possible. A power line that will provide for electrical operations of refining machinery and removal of ores from the ground will be constructed by the Alabama Power Company, according to Mr. Gardner.

Approximately 150 tons of ore will be mined when operations first begin. Production [of] up to 600 tons of ore daily will be started after opening of mining operations gets underway in December.

The removal of the precious yellow metal by the Gardner interests will be done under provisions of a ten-year lease Mr. Gardner has taken on the property by T.S. Aldrich of Birmingham.

The Hog Mountain mine was operated for a period of several decades, off and on, but production was stopped in the early days of the World War due to a scarcity of cyanide, a chemical compound made in Germany, used in extracting gold from the ores where it's found. Before the mine was closed, it had produced in its long history about $600,000 in gold bullion. This production was about three fourths of the total amount of gold taken from mines (Alabama gold mines) in the last hundred years, according to Mr. Gardner.

In its previous years of operation only about 35 tons of ore were removed daily from the veins that were worked. Because the ores therein are low grade, production of a much larger amount of ore is necessary. Therefore installation of electrical mining apparatus is to be provided for. Investigations and soundings of the property in and around the mines have been going on for the last twenty-one months. Without mishaps to intervene, production will get under way by Christmas at the latest, it is expected.

Renewed operation of the mine will be an aid to Alexander City's business houses. It again demonstrates the wealth and all-around richness of the section of which Alexander City is the trade center and capital.

Businesses in Alexander City had reason to be excited about the reopening of the Hog Mountain gold mine. Any good fortune to come out of the mine's operation would impact the economic outlook of the city. Visiting speculators, engineers and investors stayed in the old Russell Hotel in Alexander City. There was at least one prominent local investor, Benjamin Russell, who was a director on the company board with O.B. Thurlow, F.C. Weiss and R.M. Fuller, secretary. Mining experts came from other areas: P.S. Gardner, president and general manager, was from Nevada; George M. Brown, vice-president and general superintendent, from Arizona; N.O.

Surviving the Depression

Johnson, mill superintendent, from Colorado; and Elmer J. Alderfer Jr., manager of the assay office, from Colorado.

Preparations to reopen the mine began in 1931, when local men were hired to dig drainage ditches around the old sand beds and up the sides of the mountain. Heflin Cleveland, Travice Foster, Frank Smith and Frank

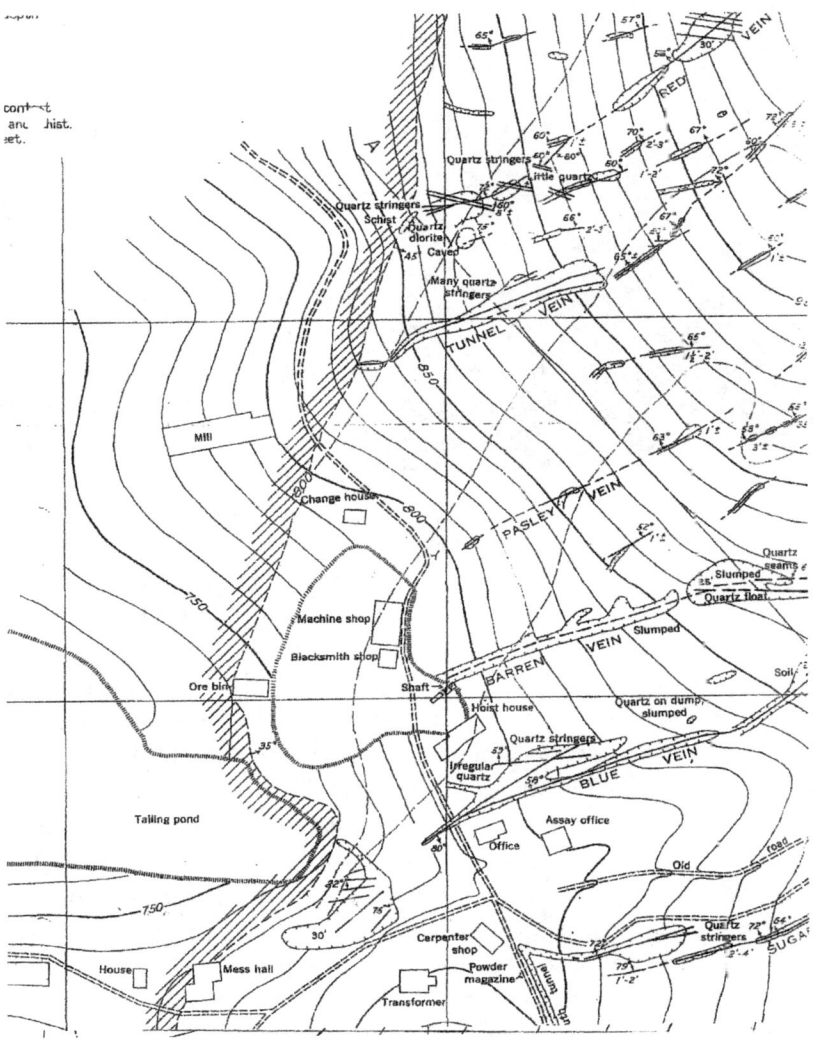

This 1948 map provides an overview of the mining structures, the machine shop, the bunk house, the mess hall, the carpenter shop and the assay office with surrounding tunnels and veins, such as the Old Tunnel Vein and the Champion Vein. *Courtesy of US Department of Interior, Geology Survey.*

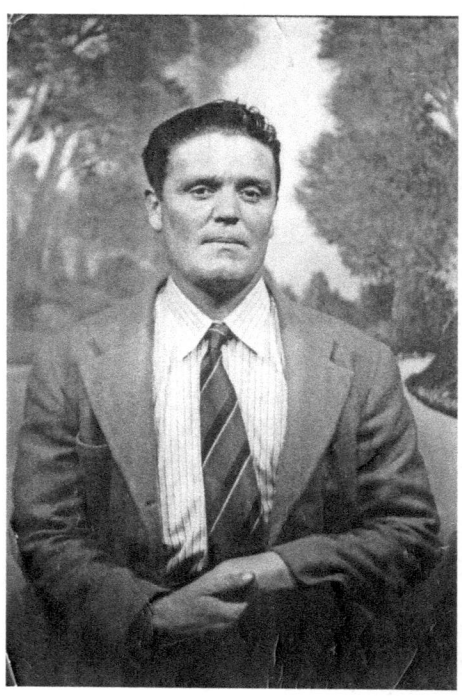

Aubrey Bowen worked as a carpenter restoring old structures at Hog Mountain Mine in the 1930s. His father, Rob Bowen, was the head carpenter. *Courtesy of Sue Bowen Smith.*

Woodruff were in the first group of men to be hired. Charlie Harris, Lander Baker and Thomas Brown used their teams of horses to smooth the sand beds. Thomas Daugherty, Llewellyn Green and a few other local men cleaned out the old north tunnel, pumping water out of the one-hundred-foot shaft. Pipes were installed as conduits for water to run into the ponds, where a cyanide and water solution was applied to extract ore from the old mine tailings. Ore samples were taken from the mine and sent to St. Louis for evaluation. Additional men were hired to dig drainage ditches leading from mine shafts flooded from underground streams. Local carpenters were recruited to repair plant buildings from the pre–World War I operation: the mill, a blacksmith shop, a bunkhouse and a mess hall.

Rob Bowen was the head carpenter. His son Aubrey Bowen, Joseph Cleveland and several other men assisted Rob in repairing not just the mill structures but also the houses on Chicken Row remaining from the pre–World War I operation. Several miners lived in the houses with their families.

After approximately two years of preparations, the mine began full production and shipping gold concentrate in 1933. The operation lasted for around four years, caused by circumstances outside the control of the mine workers. In 1937, the management began to "wind down" the mining project due to the new "wage an hour" legislation signed by President Franklin D. Roosevelt. The Fair Labor Standards Act of 1938 (FLSA) banned child labor, raised minimum hourly wages to twenty-five cents and reduced the maximum workweek to forty-four hours. One of Roosevelt's strongest supporters was Senator Hugo Black from Clay County, Alabama, also a Supreme Court justice (1937–71). Hugo

Surviving the Depression

Black was friends with US representative Lafayette Patterson from the Goldville area. Patterson and Black shared "humble beginnings" in rural northeast Tallapoosa and Clay Counties.

The story of Hog Mountain gold mining in the 1930s is told best by the old-timers who worked in and around the mine. Geological surveys, engineering reports and other official data are used to support and expand their stories. After the second period gold mining operation at Hog Mountain closed in 1916, the property remained idle until 1931.

Expert Opinions

E.H. Emerson, a mining engineer from the East, conducted research and evaluations at the Hog Mountain gold mine for a group of investors who were interested in reopening the mine. The previous operation of the mine by T.H. Aldrich Sr. and T.H. Aldrich Jr. established Hog Mountain as the top-producing gold mine in Alabama, yielding 418.44 ounces of gold.[73]

Supervising engineer N.O. Johnson observed, "The mill was started in February 1934, and, although production has been relatively small for gold mines working in this grade of ore, as compared to most other gold properties in [the] southeastern United States, production has been large."

C.F. Park observed Hog Mountain as one thousand feet above sea level and four hundred feet above the surrounding country.

> *Two rock formations have been distinguished: the Wedowee, of fine-grained dark-gray graphitic schists, and an intrusive quartz diorite of unknown age. The intrusive body, in which are the gold-bearing veins of commercial significance occupies an area about 4,800 ft. long and 800 to 1,300 ft. wide. Park described "the veins of economic importance are confined almost entirely to the quartz diorite. Toward the contact they commonly split into stringers, which pinch out."*

Preparations to reopen the gold mine were under way as early as 1931. Heflin Cleveland, his father and one brother were involved in the initial preparations. Since they lived near the mine, they had explored the old mine shafts and underground tunnels often and were eager to talk with the engineers and investors who came to inspect the mine and conduct ore testing.

Alabama Gold

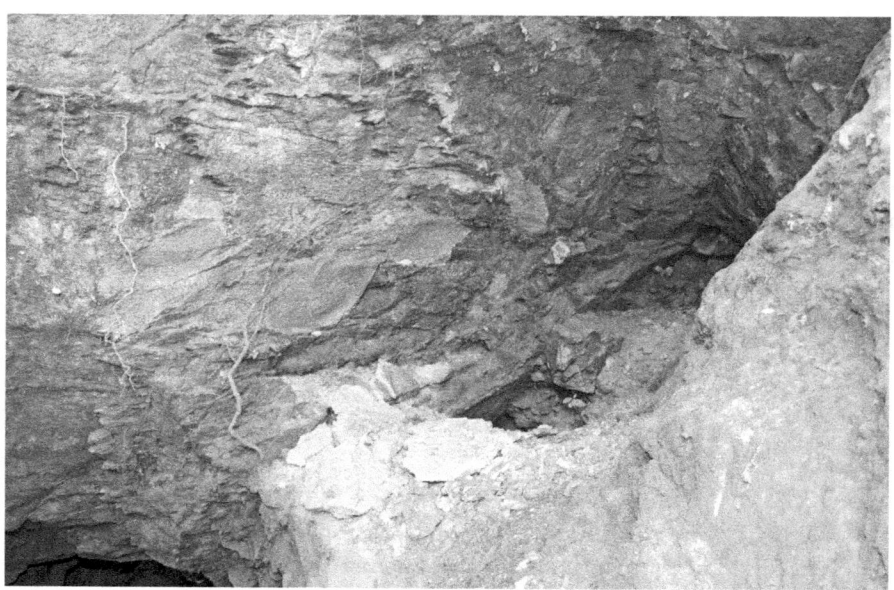

Blue Vein. Surrounding veins include Red Vein, Sugar Quartz Vein, Big Pine Vein, Little Pine Vein, Aldrich Vein and Pasley Vein. *Courtesy of Peggy Jackson Walls.*

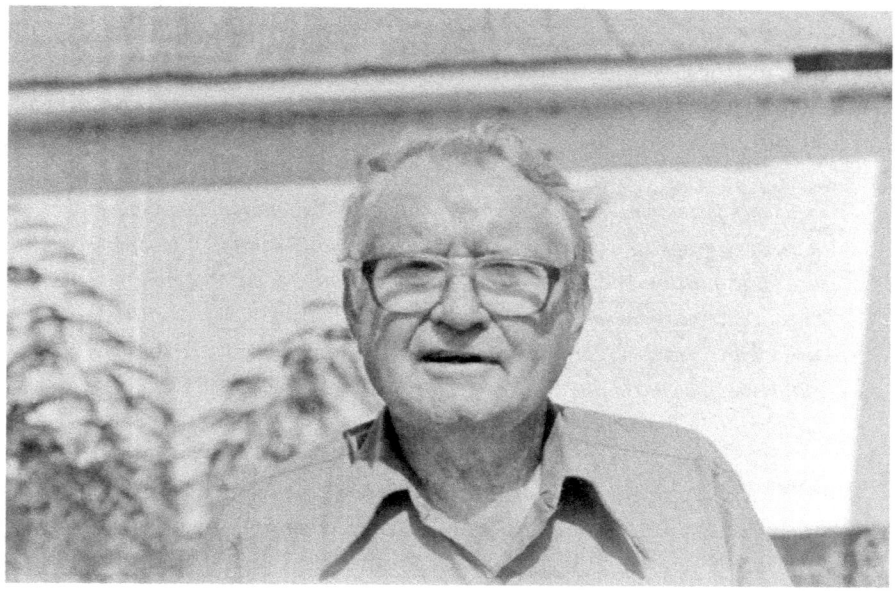

Hog Mountain miner Heflin Cleveland worked at night and went to Hackneyville High School during the day. After graduating from Auburn University, he worked for the US Army Corps of Engineers in Mobile. *Courtesy of Peggy Jackson Walls.*

Surviving the Depression

Heflin Cleveland

My brothers and I would go over to the mine into the old tunnels. These little tunnels were scattered all over the mountain. We'd go over there on Sunday lots of times and take pine knots. We would go back into the tunnels and explore them, where the old-timers had dug looking for veins. They were mining for red ore, a soft rock. Most of the red ore had been mined out, and the Tallapoosa Mining and Milling Company was looking for the blue ore that was in the quartz rock. In the last operation, the old mine had a shaft one hundred feet deep that they had drilled, and it was full of water. It had one tunnel, called the north tunnel, [that went] probably about one thousand feet back into the mountain.

At the very beginning, a man, Gardner, out of New York, was instrumental in getting the mine opened. He would get people down here investigating the possibilities. We lived about two miles from Hog Mountain. We would walk over there and listen to the men talking. They were from up north, New Jersey. We were interested in what they had to say. The owner, T.H. Aldridge Jr., was in Birmingham. People wanted to see the mine start back up. It would mean jobs for them. One of the men was Gardner, a mining engineer. There was about a year and a half of just fooling around over there before they started to work the mine.

My brother Byron worked at the mine some, and my dad, Joseph Cleveland, did carpentry work on the houses and buildings. I went to work when they were just exploring the mine, pumping out the old tunnels, drilling and testing the ore to see if it was rich enough to mine. They were working the tailings of the old mine in the sand bed that was dumped below the mine when it was in operation before World War I. They used cyanide for picking up the gold in the "tailings" of the old sand bed. The workers were running water through the sand and collecting the gold that had been left in there, and that was what I was doing when I first went to work with the mining company. I worked with Travis Foster digging ditches with a pick and a shovel around the sand bed for drainage so no water could come in from the mountain. We started at the top and dug a ditch at least two feet wide and from two to three feet deep to catch all the water. I had to be on twelve hours, and Travice had to be on twelve hours. We relieved each other. One of us to be there at all times.

We had to dig ditches down to the solid ground in the sand bed. First, they'd pump lime water to neutralize it and pump cyanide water into these ponds and let it soak through the sand and pick up the gold. Then, they would pump it up the hill into a tank that had zinc shavings in it. The zinc shavings would dissolve and react with the cyanide to pick up the gold and settle it out as a solid. Black mud is what it looked like before it was shipped off. Ed Walls looked after those tanks that had zinc shavings in it in that operation. This was when I first went to work.

After they got the water pumped out of the shaft and got it cleared, they brought in one driller from Birmingham. He drilled and extended the tunnel some, looking for veins. Then, they brought in mining engineers who took samples to determine how much gold there was. Finally, they decided mining was worthwhile, and they sold stock in the mine, mostly to people up North. The main shaft was one hundred feet deep from the 1890 operation. They drilled down one hundred feet and ran two, short tunnels off from it. Then they run this north tunnel up into the mountain. It started at ground level when they first pumped it out and went up. It was not over two or three hundred feet back into the mountain when they got through. When they quit, it was probably a quarter of a mile long or longer. Usually in running those tunnels, they had somebody working all the time on the two-hundred-foot level in the north tunnel. They would trail out and shoot about five feet in a shift. They drilled it, shot it and hauled it out. They would make a five-foot progress in a shift. A shift was ten hours because they were running just one shift a day for about a year. They didn't open up the two shifts until they got the mill crushing and running the ore.

I went to working down in the mines when they first started what they called "mucking." The drillers would drill spaces in the tunnels and shoot the rock down. Then the muckers would shovel it up and throw it into cars. The tram cars hauled the rock out, ran it to the shaft and put it on the elevator. They'd pull it up, roll it out and dump it. This was just blue granite rock; it wasn't ore. They were just looking for the veins to get ore from. I worked in mucking for a while, but I didn't stay long before I got a job on top. I didn't like it in the mine; there was always powder smoke from shooting, and it gave me a terrific headache nearly every time I went down there.

I worked two hundred feet straight down. It was already one hundred feet with two tunnels going off the main. They drilled another hundred feet and ran off tunnels under it. Then when they used blowing charges, they found the veins; they lay slanting up, all the way up in layers between blue granite rocks. We'd start shooting that stuff out, and they might go fifty or sixty feet blasting with dynamite. J.P. Mooney and Roy Mooney were in charge. They learned about blasting from the other people at New Site. Frank and Mon Woodruff operated drills over there. I helped

Surviving the Depression

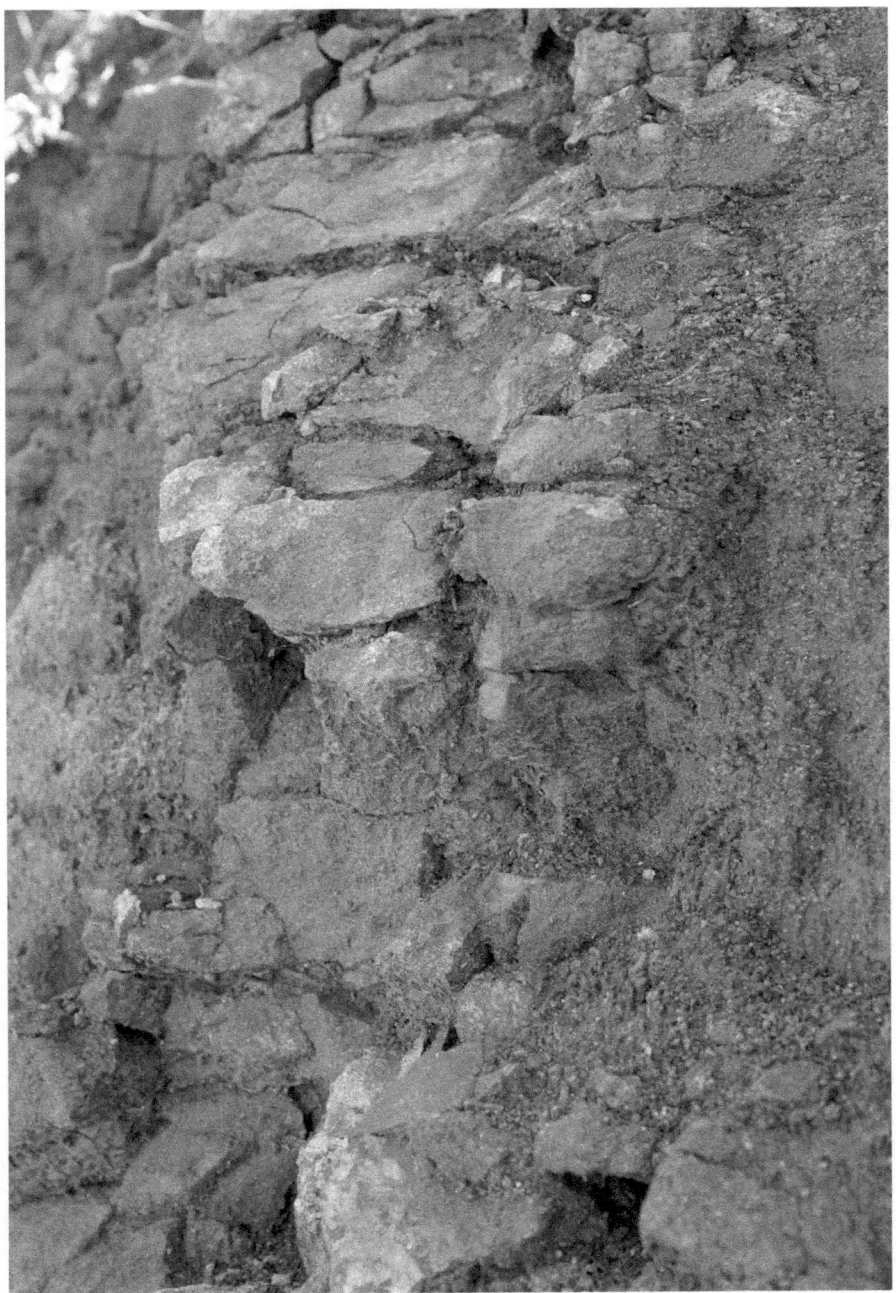

This quartz vein was the location of several gold deposits. More gold has been produced at the Hog Mountain District than any other location in Alabama. *Courtesy of Peggy Jackson Walls.*

Mon on the drill some, and we'd drilled up—what they called "drilling the face out." Then we'd take the powder, stamp it in and shoot it. Before we got the holes loaded, we had to move our timbers out so the rocks wouldn't be shot down on the workers. All the men who worked down in the mine for any time developed silicosis—Thomas Daughtery, (Rufus), Kermit Jackson, Frank and Mon Woodruff, Cecil Osborn. They worked in the mine longer than I did.

I went to work on top in the machine shop, helped in the pipefitting. Carey White was the chief machinist and pipefitter there. Richard White and his daddy were blacksmiths. Carey White helped put in most all of the pipe and pipefitter. I worked with him some in the mine. You had to run pipe for the air compressors to pump air. All the drills run compressed air. You had to run pipes down to keep it running into the head so that as fast as they extended the tunnel, they had to run the pipe on. They put air down into the drill and water. They worked on the same principle as a jackhammer. Only they were better and heavier drills that were fastened on a beam. When they drilled, it made fine dirt, and they used water to blow it out. So you were always wet in there, and that drill made a terrific noise. In fact, I don't hear good now from working down there with those drills. When you first came out of the mine, you couldn't hear anything. You couldn't talk down there 'cause the noise was so bad.

That was continuous the whole time you were down there. The drilling shifts run for nine hours. They'd go in and set the drills up. They had a stand that they would wedge from the top of the shaft to the bottom, and they would screw it into where it was perfectly tight about two or three feet from the face of the tunnel. Then they'd put the drill on it, and they could move it up and down or cross sideways. They drilled about eighteen holes anywhere from six feet deep. They'd start drilling them kind of in a wedge shape so whenever they shot, they'd shoot out a place about three to four feet deep each time.

Then the muckers would come in, shovel and put the rock[s] in cars, roll them out and dump them outside. The man [who] ran the elevator brought it up. The elevator sat to the side. When the man loaded the car, he'd ring the bell and give a signal. They had different rings for different signals. When the car reached the top, the top-house man would roll it off to the dump if it was waste, if it was rock. If it was ore, he'd run it out to the ore bin and dump it in the bin. The crusher was down at the bottom of this ore bin, where ore went into the rock crusher, where large pieces were crushed into small size rock. It went down on a conveyor belt, and this

Surviving the Depression

conveyor belt carried it up the hill to the ball mill that ground it into fine powder. Then it went into these vats where it was mixed with chemicals and agitated like something similar to a cream separator. The ore was heavier; the gold was heavier than the rock just like cream is heavier than milk. Well it felt like black mud. The gold was in a chemical mixture. I believe when the ore was mined out, the richest of it was about thirteen dollars to the ton of rock. I think that was some of the richest ore they had over there—anything from that on down—and when they got through with it at the ball mill, this concentrated stuff that they would send off, they'd send it off to a smelting plant, where they took it and really got the gold out of it. They didn't do that at Hog Mountain.

When the mine first opened, most of the labor was done by local people. And in the process, some of them learned to do skilled jobs. Other skilled and experienced workers came from other mines. There was a fellar [named] Proctor in Birmingham, Mr. Luther Dye and Mr. Davis. George Brown was the superintendent over the whole thing. He was a graduate of the Denver Mining School. Mr. Biggs was a chemical engineer. And there was a civil engineer, who was the one [who] took care of keeping the tunnels going in the right direction and overseeing that they went where they were supposed to go. He ran a transit and kept those tunnels lined up and leveled. Of course, the drillers had to help keep the tunnels level, too, but he was the one that kept things start.

When I was working over there, I was making $1.25 to $1.75 a day working twelve hours a day. The main shift ran eight hours, but the ball mill ran eight-hour shifts because they had three people to look after them. Buildings included an assay office where they could assay their own samples. They built a big blacksmith shop where they could sharpen their drills. They put up a rock crusher at the ball mill where they crushed rock to a small size like gravel. They carried it up the hill on a conveyor belt, and the ball mill would grind it just as fine as dirt. That's what it looked like—black dirt. They'd run some kind of chemicals to pick up the gold and float it to the top, where they skimmed it off, like skimming cream off milk. But they'd get some of the fine dirt in the gold. They squeeze it out, put it into bags and send it off. The rest would be dumped. The ball mill didn't shut down unless they had a breakdown.

In 1936, Heflin left Hog Mountain Mining and Milling Company and worked three years for Russell Manufacturing Company in nearby Alexander City. Then, he worked for Alabama Power Company until World War II began. After serving in the US

Alabama Gold

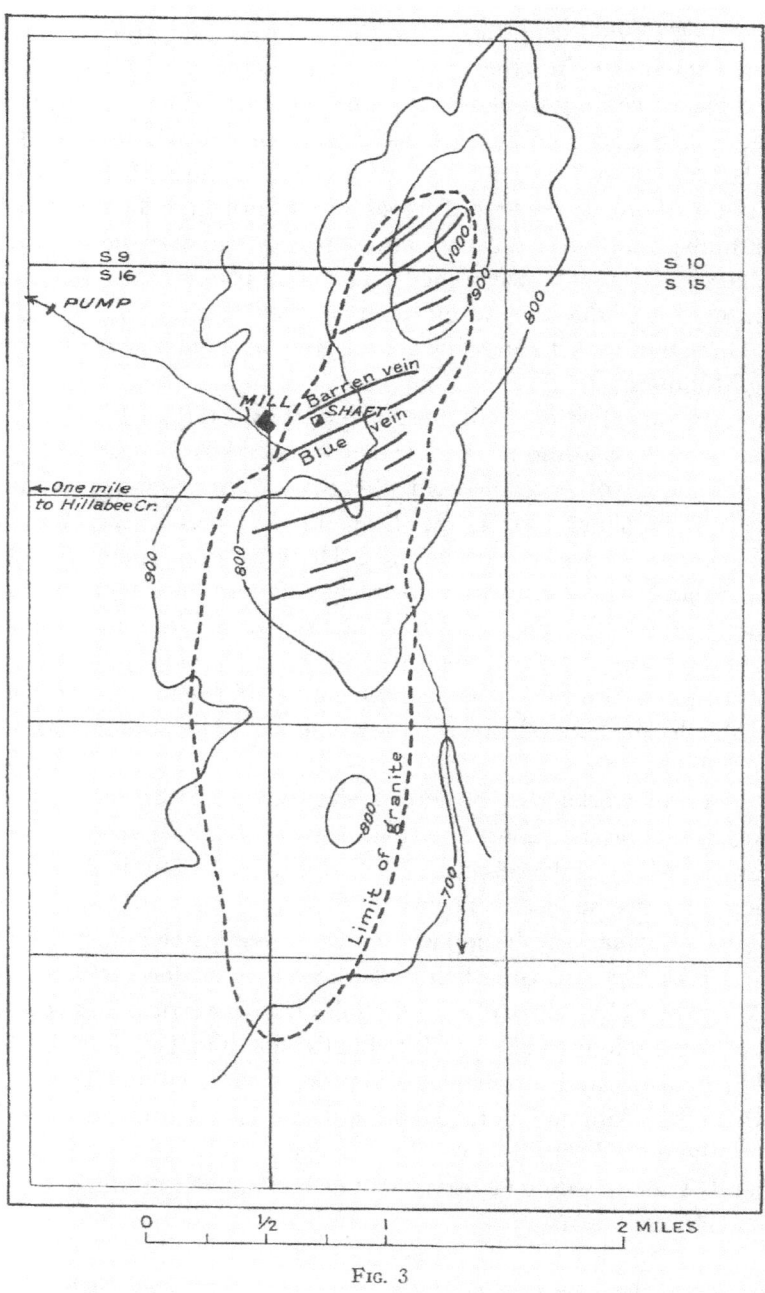

Tallapoosa gold mine map. *George I. Adams, "Geological Survey of Alabama,"* Bulletin *40, Fig. 3, 48; courtesy of the Alabama Department of Archives and History Digital Archives.*

Surviving the Depression

Army for four years, Heflin entered Troy State University as an engineering student. After completing an agricultural engineering degree at Auburn in 1953, he worked for the US Army Corps of Engineers in the Mobile District until he retired.

Llewellyn Green

Llewellyn Green went to work at Hog Mountain in 1931 when Tallapoosa Mining and Milling Company was making preparations to reopen the mine and was involved with different aspects of work. Three of his brothers—Carl, Willie and Wallace—later worked at the mine. His mother, Miss Eula, and his father, Zack Green, took in boarders. With a touch of Old Southwest humor, Llewellyn tells his story of working at Hog Mountain in the 1930s.

In '20s, my daddy bought an old '25 T-Model Ford truck for $50.00 and two cords of wood. We moved people and any kind of hauling from cattle to logs. I started driving the truck when I was fifteen years old and was helping on the family farm when I married in 1931. During that winter,

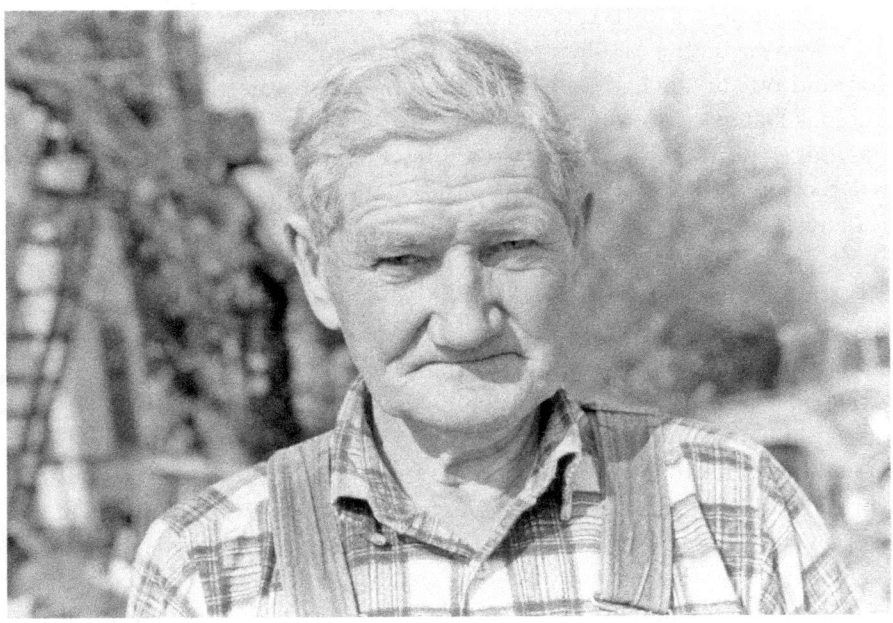

Hog Mountain miner Llewellyn Green was "one of the first three men hired at the mine and one of the last three to leave." *Courtesy of Peggy Jackson Walls.*

Alabama Gold

Prentice Foshee and Leon Watley were looking after the mining project for Mr. Aldrich. So we three went to work over there—to shovel dirt and rock for samples to send to Mr. P.S. Gardner and Mr. Aldridge "up at New York." After that more workers were hired, and the project was in full operation in 1932. The operation grew to [be] "bigger and better," but at the same time, it didn't pay very much money. I worked there for 10.5 cents an hour, twelve hours for $1.25—for $8.75 for a day's work. That was seven days a week on a twelve-hour shift. I had to walk three miles to work and three back after this twelve hours of work each day.

I done everything from firing the boiler to cleaning out the shaft and hauling rock and ore on a little ole car. I taken samples to the assay office and helped fix foundations for buildings before we got to mining ore. The mine picked up, and the management got people in Alexander City to take stock in the mine. They used First National Bank, Mr. Russell's bank, to put the stockholders' money in. That's where we got our little pay in an envelope, in cash money, each week. Mr. George Brown was the mine superintendent; George Biggs was the mining boss, and I worked in the mine under him. Then Mr. Session was a survey man, and Mr. Simons was an assay man. I worked under Neil Johnson, the mine foreman when we got the mine started. I worked down in the mine. Simon and Cowley sunk this shaft down from the 100-foot level [to] another 125 feet. The old buildings from old operations had fallen into the shaft, and water was in there, too. Thomas Daughtery and I tried to get out those old timbers that were hung and lodged down in the 100-foot shaft. It was hard, nasty work, but we were proud of our 10.5 cents an hour.

Later, Thomas and I operated the hoisting engine to raise the cage that carried men up and down. We used it to bring up rock, too. Some others who worked there were my brother Willie. We worked together in the assay office and then my brother Carl [and] Thomas Daughtery, Hardy Buckner, Frank Bowen, Tommy Brown was a night watchman there, Morgan Dean, X Dean, Clark Dean, Talmadge and Kermit Jackson, Rhett McWright, James William Steward and William Worthy. Everybody was trained and knew how to do their job. Everybody was very lucky to have done all the work we did and not have some serious accidents in the mine. Among all the people who worked there and could take samples, I could have taken some, but if I had, I didn't have a way to get rid of it without a record of where it came from. I could have been a wealthy man because I knew more about where the free gold was in all this washing and traveling of the sand, water and ore in what we called the "tailings."

Surviving the Depression

I worked on until I was among the last three [who] left there in '38. The mines had to shut down due to the fact that it cost so much to mine this ore and separate the gold from the granite. The stockholders began to think they were going to lose what they had in it, and all of them began to pull out. They told everybody working there, "If you've got another chance to work, you'd better take it because the mine is gonna be shut down."

Then you had to eat what you could get and that was it. We sold chickens for ten cents a piece that weighted a pound and a half. Sold cotton five cents a pound. Worked all day for one gallon of sorghum syrup. Could get three gallons for a dollar if I had the dollar, but I didn't, so I worked from daylight to dark for one gallon of sorghum syrup. Furnished my own lard bucket to put it in. Worked for a bushel of corn for forty cents a day as hard as I could to help gather a man's crop—spare time work between the mine and farm work. When the mine opened up, it was a boon for Tallapoosa County at that time. People got employment and made more money there than they could on the farm.

At the present time, it's doubtful that the mine could be opened and worked for a profit. There's lots of gold there, but per ton, it's not all that valuable. Some of it is good value. Anyhow in this north tunnel, I drilled into a vein eleven feet wide, and my boss man told me [to] "go ahead and drill a heading in the foot wall and cover this thing up." So me being young

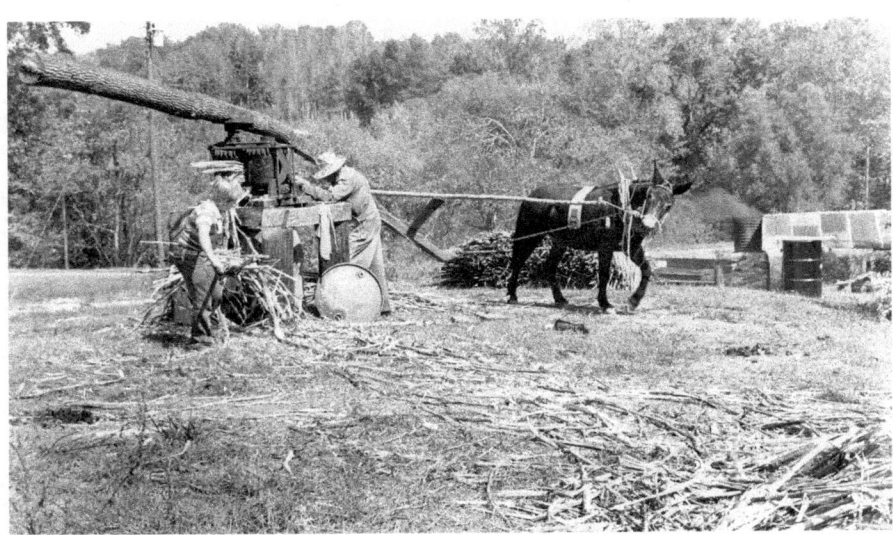

John Johnson and his son Ray operate an old-fashioned syrup mill on John and Judy Kendrick's property in 1983. *Courtesy of Peggy Jackson Walls.*

and interested in things, I asked him, "Why should I cover this up?" He said, "Well, in a day to come, at the present time of wages going up, this mine might not run too long and later on, we might operate this mine ourselves." Well, after things got to be a little more expensive, the stockholders didn't get quite as much money out of the gold for their stockholder, and they began to pull out. That's how come the mine shut down. But if wages get back down again where people won't want a fortune to work, it will run again. There's lots of gold there.

Llewellyn married in 1931, when he first went to work for the Hog Mountain Mining and Milling Company. His wife, Vida, came from a family of miners and contributed information about her family's involvement in the Hog Mountain Mining and Milling operation in the 1930s.

Vida Mae Bonner Green

My daddy, Bennie H. Bonner, worked at the mine, carpentering, but he didn't work all the time. Clarence Butler, my first husband, was the night watchman. My two brothers, John and Hershel Bonner, worked at Hog Mountain, too. Hershel, the youngest, was working over there when the mine shut down. He went with Mr. Brown to North Carolina to work there. My daddy and my husband helped to build a large slab house that Mr. Brown lived in during the operation of the mine. Anybody would think of slabs as something to throw away, but the way they fixed it was real beautiful. They had a bunkhouse and a mess hall where the men stayed [who] lived too far way to go back and forth to their work. They had a cook, Mr. Frank Rowan, and he lived close to the mine in a house on Chicken Row. The houses were from the earlier operation that ended in 1916. There were houses on both sides of the mountain that people lived in. A lot of people lived there. People didn't make much money, but the men stayed in the bunkhouse and ate in the mess hall. They taken so much a week board on 'em while they stayed there. I believe it was a dollar and a half a week.

Some miners brought food; others ate in the mess hall. There were rolling stores that came by once or twice a week. Marvin Allen and Hester Eason toured the community and went by the mine. They had canned goods out of the stores: dried peas, butterbeans, sweet potatoes, Irish potatoes, cabbage and vegetables in season. Mr. Allen brought vegetables in jars that his wife canned.

Surviving the Depression

Thomas Brown

In addition to his own mining experiences in the 1930s, Thomas Brown had several family members who worked at Hog Mountain mine. An older brother, Doc, worked at the Hillabee gold mine operation and shared stories with Thomas about the Hillabee gold mine operation. His ancestors lived near Hog Mountain all the way back to pioneer days.

My father was William Alexander Brown. His father came from Spartanburg, South Carolina. He ran a blacksmith shop and a ferry and built a house in 1854 on this land, where I was born and raised. I had five sisters, two brothers and then two half sisters by my father's second wife. I finished seven grades at Valley Grove School, finished high school at New Site [and] then wound up my education at Massey Business College in Birmingham in 1922.

The mines operated two different times. Now I know when the mines run the first time [pre–World War I]; they built a dam over there on Enitachopco. They made their own electricity, and operated it at the mine. Jim Britton was the fellow [who] stayed there and looked after the power plant, fired the boiler and all. There was a superintendent whose name was Kennedy. He

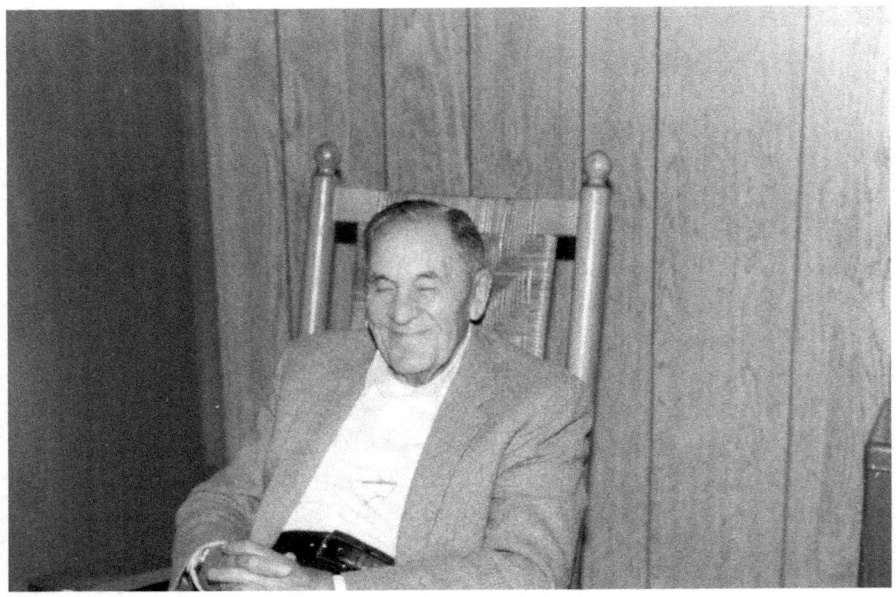

Hog Mountain miner Thomas Brown helped to clear the mining site in 1931 with his team of horses. He operated the hoisting machine. *Courtesy of Peggy Jackson Walls.*

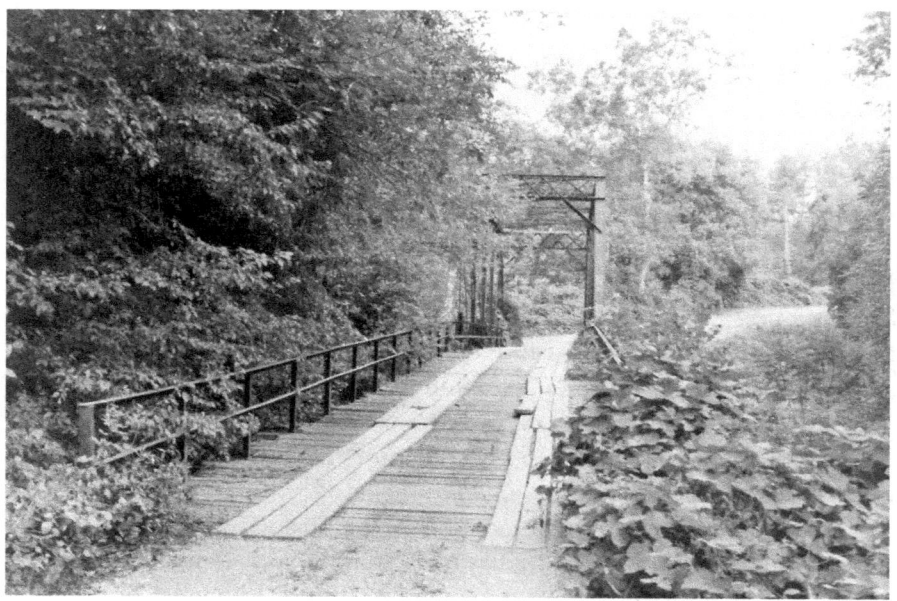

The Enitachopco Creek Bridge and dirt road led to Hog Mountain mine. *Courtesy of Peggy Jackson Walls.*

was a smart man, but he went over there and was on the pond in a boat and went over the dam and got killed.

I worked there before the mining got underway 'cause I had a team of horses. George Brown, the superintendent, asked me to take my team up there and haul sand to build foundations for the machinery. So I went over there and hauled sand. Three or four other wagons hauled sand: Lander Baker, ole man John Harris, Frank Downs and myself. I didn't work for about a year after we got through hauling the sand. Then the superintendent of the mill came one night and wanted to know if I would take a job over there running the drier. I went to work then, dried the concentrate after it came out of the flotation machine, and worked until the mines shut down.

I worked at the ball mill, where the ore came from the crushers. It wasn't anything unusual to pull two hundred tons of rock out of there in eight hours. They had two crushers that ground the rocks down small; then they had one that cut it down very small. The ball mill up on the hill was filled with steel balls, rolling grind ore just like flour. It came out at the ball mill [and went] then into a trough of water and went down to a flotation machine. The water would go out, and this concentrate would float and go into a big tank; then it would come out of the tank into a big ole thing like a boiler except

Surviving the Depression

that it was rolling. It was steam that dried the stuff. When it came out on the concrete floor, it was cut into samples and sent to the assay office to see how much gold was running a ton.

I worked as supply man for a while. They had a big board, had nails drove up, and they was numbered. Each of the workers had dog tags with a number. When somebody come in to work in the evening, if he was assigned to the old north tunnel or to the eleventh stope, he'd hang his number up there at that name. Then when a shift ended, if anybody was missing, they could look up on that board and tell where to go to find them. Well, there was over two hundred names up there. We did have to look for them a time or two. I remember one Saturday night, wasn't anybody working in the mines on Saturday night but the timbermen and the plumbers. Of course, they had these water pipes in there to keep these drills wet. And the plumbers was in there.

When the day shift went off, they'd always shot the holes they had drilled, they'd shoot the ore down. Well after they'd shot these holes, they waited for the smoke to get out before they went in. Ben Buckner went into a place where this damp air was, and it got him. I was listening for the bell for the hoist; and after a while, they give me a signal to pull a man out. I pulled the hoist up, and they brought him out. He didn't know nothing, knocked him out. I went to the office and got some ammonia. He was panting like a bilious horse trying to breathe. We got that ammonia to him, straightened him out, and he went back in the mine before the day.

But dynamite smoke was the biggest problem they had over there. I used to go down in the mines lots of times in the mornings, when the night shift was getting off, just to hear them shoot. They had those shots timed. Maybe this driller here had fifty holes drilled in the wall, and he'd have fifty loads in there to shoot. This man rigged the shots; he timed those things with his fuses, where they was just like a clock ticking going off. It was interesting. You might be standing a half mile down the tunnel when the shooting started—it'd blow your light out, [the] air coming down that tunnel so.

Dr. Street was the company doctor. I carried several men down there in my Model A that got banged up a bit. The man [who] did the night watching had high blood pressure, and the company doctor stopped him from watching at night 'cause it was dangerous. He was liable to fall in the machinery or something and get killed. Mill superintendent asked me one night if I'd take that job, and I did. I stayed on it until the mine shut down.

I worked over there from '32 to 1937, when the mine closed. A day or two before I quit, Mr. Brown called me into his office and told me, "The mine is fixing to close down. If you can get you a job somewhere else, go ahead, and all you've got to do is just come by and tell me." Well, I wasn't thinking about getting a job. I had a farm here, two mules and plenty of work to do. I wasn't worried about it. But two or three days after that, they all came into work one night, and Brown got out there at the shaft where they'd all gathered to go down in the mines, and he told them all that the mines was closing down. If they could find work somewhere else to go ahead get 'em a job. A geologist, Frank Feyer, from Colorado told Mr. Brown that if they would sink that shaft over there another hundred feet, they would find the richest gold they'd ever found. The superintendent George Brown asked for the money to sink the shaft, but the stockholders didn't grant it. Mr. Brown left here and went to a mine in Tennessee.

Hugh Price McLeod

I was born near Valley Grove, up around Hog Mountain in 1912. My parents were Mr. and Mrs. C.B. McLeod. They were sharecroppers. My father and father-in-law worked for the Hillabee Gold Mining Company. My wife's brother, Madison Jones, also worked for Tallapoosa Mining and Milling Company. I worked there five years [and] then went to work for Russell Manufacturing Company in 1936.

My first job was helping to clean out the one-hundred-foot-deep shaft and pump the water out. My job was cleaning out in the tunnels, so they could sink the shaft another one hundred feet. The miners knew where the veins lay in there, and they claimed that most of the time, the deeper you'd go, the richer the veins would get. They also claimed that you can get down under those veins, laying in a forty-five-degree angle. They run northeast and southwest. The miners would go down with their shaft and cut across these veins, and when they'd strike the vein, then they'd follow it in the tunnel, where they would put the track for the cars to run on and haul out ore. When they made a strike, they'd follow the vein, putting down track to run the cars on. Kermit Jackson built the chutes where they pulled the ore down and shoveled it up. They'd push this car up under there and pull it loose. They could just fill that car up in no time flat. Where, if a man was a-shoveling it off the tunnel floor, it would take him a long time to load one of them cars. And when he loaded it, he'd take it down to the elevator and then we drawed it out.

Later, they moved me out from there and put me in the drill shop, heating steel in a clay brick furnace. The furnace could be fueled with kerosene, coal

Surviving the Depression

oil or fuel oil. We sharpened the steel that they used in drills. I ran a pipe into the furnace and opened a valve, lit it and connected it with the air pipe. It was awful hot air that blowed out of that pipe. It had to be to heat steel to where they could sharpen it. The drill sharpener was run by air. After the steel was sharpened, we tempered it. After we got through with them, those drills were sharp enough to drill holes in solid rock. Some of the men I worked with in the drill shop at different times were Bill Whiting, Carey White, Llewellyn Green, Hershel Peppers and Thomas Daugherty.

When I first went to work in the drill shop, I worked with Carey White. I recall when Carey lost his finger in an accident down in the mine shaft. Carey was the one in charge of keeping the drills in order, and he was responsible for the pump that kept water pumped out of the mine. It was like it was raining down there all the time with water coming out of the rocks. Well the power was off one day. Carey and a helper went down to work on the pump. After Carey packed the pump, he got up to start out and slipped and fell. His little finger went right straddle this rod that the pump run on, and it clipped his little finger off all except a little piece of skin holding it. He turned to his buddy and told him to cut it off. The fellow that was with him got his knife, finished cutting it off and put it in his pocket. There wasn't any power, so Carey had to climb the two hundred feet up the ladder. His helper told him, "You go first. If you fall, maybe I'll catch you." We had a good laugh despite the seriousness of the situation.

We did like to have a bad accident sure enough one night when the hoist man went to sleep on the job. J.P. Mooney…we called him the powder monkey 'cause he was in charge of fixing up all the dynamite to shoot the ore out. Well, it was the third shift, and J.P. Mooney got on the elevator with about three cases of dynamite and a bunch of electrical caps. J.P. signaled Thomas to pull him out. When the elevator got to where J.P. was supposed to get off, Thomas was asleep. The hoisting engine pulled the elevator up to the top of the head frame, where it could go no further. It pulled the elevator over enough so that the wire that was fastened to it broke and stopped the motor. Nobody was hurt, but if that bunch of electrical caps had gone off, the whole place would have been blown to kingdom come.

When I worked in the drill shop, there was just two shifts. But when I worked in the mill, I worked the swing shift, and I worked with different men on each shift. The swing shifts run eight hours. There were three eight-hour shifts, and we would change every two weeks. We would work the first shift two weeks, then the second two weeks and the third two weeks. The mine run two nine-hour shifts. We got paid every two weeks. The most I got was

Alabama Gold

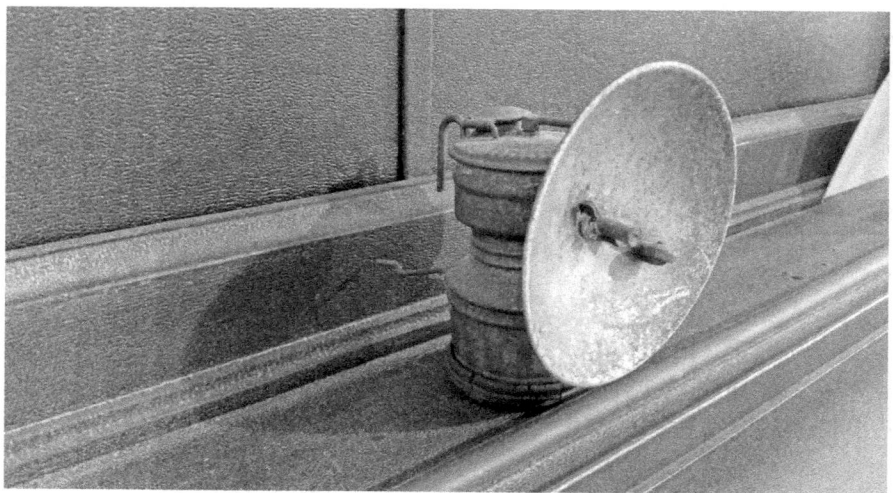

A carbide lamp could emit bright light in the mines. The lamp also could be used to ignite fuses dynamite fuses. *Courtesy of Tallapoosee County Historical Museum.*

$1.75 a day. The man [who] run the machine for sharpening drills made top pay—$3.00 per day. I walked back and forth every day 'til I got enough money to buy an old '29 A Model.

Soon after I went to work there, one of the first things they did was clear out a right of way to run the line in there for the machinery. Alabama Power set up special lines to the mine before there was electricity in the surrounding communities. The equipment was brought in a big truck. Before electricity, there was only the carbide light they used in the mine. Those things made a pretty good light down under the ground. See they got a little reflector on them. You light that and put it in a dark place. As dark as it was down underground, it made a good light to work by, enough that people could see how to do their job. They didn't never have electricity down in the mines. There wasn't never no power down in there.

Blasting was the last thing they did on a shift. The drillers would go down and set up. They'd drill out what they called a heading, what it takes them, they had to drill several holes, maybe six, seven or eight feet. I believe six-foot steel was the longest we had. They'd start with a short piece of steel and drill that as far as it would go. Then they'd put in a three-foot piece and on like that 'til they go to six feet. It would take from eight to twelve holes in what they called the heading. They would load all those holes with dynamite, and it would take 'em the whole shift to drill that many in solid rock. It was a slow process. What I mean, the drill didn't go down like it would out in the

Surviving the Depression

dirt. Anyway, it would take that long for them to drill out what they called a heading and then load it. They had to pack these sticks of dynamite in the holes, ever how many sticks they thought it would take. At the end of the shift, that's the last thing they did. They would light those fuses and then go out before they burnt to the cap to blast it off. They would have time to go from wherever they were to out of danger. When the next shift came on, they had to wait thirty or forty minutes to get that smoke blown out. They had fans up there at the top down in the mine to blow that smoke out. If it didn't, a person couldn't work in that dynamite smoke.

It [took] six hands all the time around the clock to run the mill, two to a shift. When I worked at the mill, I run the ball mill part, and Ed Walls run the flotation machine. The ball mill was where the ore came out of the bin after it going through the crusher. What I had to do was to take a sample of that stuff at six or eight places every thirty minutes. We had a sample coming in and one going out. I'd get a sample and put it on a piece of paper. Someone else carried it to the assay office, where they could know how the ore was running in mineral value all the time.

The superintendent Mr. Brown came in from Birmingham. They had a stockholders' meeting. He came on out to the shop, where we were making up some new steel. Steel would come in great, long lengths, and we were cutting it up. He said the stockholders voted to mine up what they had shot down and call it quits. And that's what they did. They claimed they had mined out when they sunk the shaft the other one hundred feet after they opened up the second time. They had a lot of ore that was shot down in these stokes. There was still a lot they hadn't loaded out to be mined, and they had got about all that they could from this other level. They had got it all to the first level. They claimed it was going to cost so much to sink the shaft under this ore to get it. They claimed everything was going up with the eight-hour law and wages. They decided it was going to cost more than they could afford to get it out. This was in 1936, but they run on. I believe it was in 1937 when they shut down.

The only other type of work in the community was farming and sawmilling. I think most of the farming was sharecropping. Most of 'em done it on halves, you know—the men [who] owned the farms would furnish everything—mules and tools. A lot of 'em worked on a fourth, when the men [who] was farming owned their own mule and tools. And where it was on halves, the men [who] owned the land furnished everything for the men to work with. Then, at the end of the year at gather time, he would get half of what the farmer made, or fourth, whichever way it was arranged.

Alabama Gold

Born in 1912 at Valley Grove in the Hog Mountain District, Price left school in the eighth grade to work in the fields with his father. His mother also contributed to the family income. "She'd raise chickens and sell eggs to buy what little things she had to have like sugar and coffee, and other than that, we didn't have no money…didn't have nothing, only a lot of love, I reckon." Taking care of family was foremost in Price's life whether the money was earned from sharecropping, mining or millwork. In 1937, Price left the mine and went to work for Russell Mills in Alexander City, where he remained for thirty-eight years before retiring.

Marshal Edwin Walls

I started off working on the sand beds at twelve hours a day and night. The hours were from six in the evening to six in the morning. They took the sand bed and built what we call pens, ponds like rice paddies, only smaller, twenty to thirty feet around and built up about knee high. They pumped cyanide solution in the pens. It would run down through to a ditch they had cut.

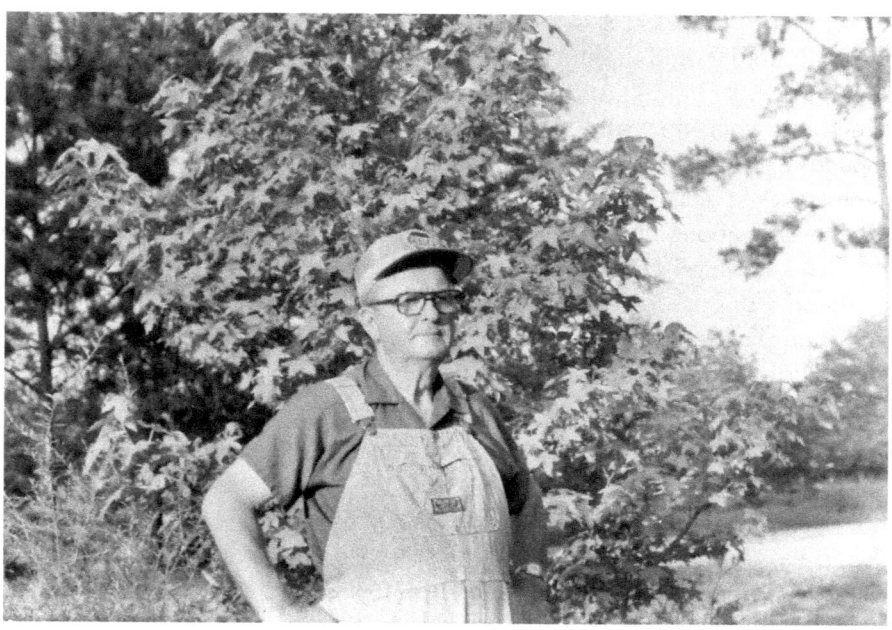

Marshall Edwin Walls was a Hog Mountain miner in the 1930s who mixed chemicals for processing ore and operated the flotation machine. *Courtesy of Peggy Jackson Walls.*

Surviving the Depression

Down at the lower end of the ditch they run it in a vat of zinc to recover the gold out of the cyanide solution. Free gold, or nuggets, we'd run through a sluice box or pan. Most of it was run through a sluice box and recovered on velvet. It settled to the bottom, and the sand and stuff would wash on over. We would take the velvet out, empty it and get the free gold out. That's the gold that's not in the compound. The nuggets were big as a pinhead on up; that's what you recover in a sluice box. The old tailings they couldn't recover in a sluice box, you pour the cyanide solution in it; then [you] recover it with zinc powder. And it is deadly poisonous. The zinc gathered in the ore like iron fillings going to a magnet. Then when we got so much of it, it would be canned up and mailed to Long Island to a smelter. When they smelt it, the zinc passed off as gas and left the gold. The company developed a new method of separating gold from the compound, mostly in an iron oxide.

To get ore out of the mine, they would tunnel under the vein, shoot it down and haul it out to the elevator. Then they would bring it up in the little tramcars. They dumped it into the ore bin, run it through a jaw crusher and then the gyro-crusher. From there, it went to the ball mill. There was just one jaw crusher. They crushed up the big rocks to one to three inches. Then they went up on a belt and were poured off in another ore bin, and when it was going on that belt, they had boys on each side picking out the granite, just racking it off. It went down into the waste dump. They didn't have much but ore going into the bin. The ore would go through a gyro-crusher which would crust it up into about a quarter-inch.

The gyro went round and round like a colander. It had an eccentric at the bottom where it went around inside of a bell. The crusher pulverized the ore into an eighth to a quarter inch. Then the ore went up on a belt two hundred and something feet, maybe up to five hundred feet. Then poured off into the bin and from there to the ball mill. The ball mill was a big drum that turned. It was about fifteen feet high and about eight feet wide and turned, and they loaded that with big balls anywhere from one to four inches...and that would crush it up into about three hundred mash, gold dust. That's about twice as fine as flour.

They went from the ball mill to another machine, like a sluice box. The heavy sand went down under, and the fine sand floated to the top, and had returns for the coarse sand to go back into the ball mill and be reground until it got fine enough. Nothing but the two hundred mash that went from there to the flotation machine, where I mixed the chemicals and free agents to catch the gold. They had an oily substance and a powder. You mixed so much with the solution, and it went through. The flotation machine turned

Alabama Gold

about two thousand revolutions a minute. It had augers made out of hard rubber that sucked the stuff up and mixed it all together. Then it floated over the top and ran off down into a trough to the drier. It was dried down there and canned up, all the minerals together to be shipped to the smelter in Long Island, New York. That smelting heat was four or five thousand degrees. When they smelted it, it was twenty-four-karat gold.

There were about two hundred people who worked at Hog Mountain. Inside the mine, there were drillers, muckers and loaders. They loaded the boxcars and brought them down to the elevator and brought 'em out. On the top of the ground at the elevator, there were two top-house boys who carried the cars from the elevator to the top of the ore bin and dumped ore into the bin. There were hoisters, machinists and mechanics at the machine shop. Several men worked there, and then they had a drill shop and a helper on three shifts—carpenters, electricians and general flunkies. They came from all over. Some of them boarded in the bunkhouse. We had a mess hall, a regular crew of helpers. There were thirty or forty [who] boarded at the bunkhouse. Then there were several families that lived in the houses down on Chicken Row, where the road comes out.

There was one huge building, the mill building, where the refining process was. And the crusher house, it was three stories high. It was down the hill,

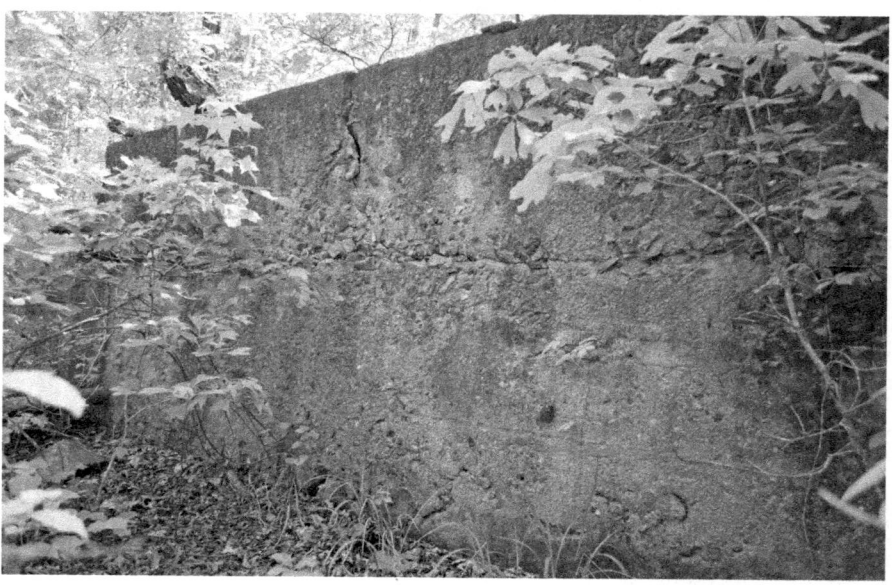

These remains from the 1930s Hog Mountain mining operation are the part of the old mill wall. *Courtesy of Peggy Jackson Walls.*

Surviving the Depression

like steps. That was where they crushed it up, and it went from there to the mill on the hill, the ball mill. And the office building and the boss's dwelling house over the hill. They had a miniature smelter there for the samples, the bunkhouse and the dwelling houses on Chicken Row. They served three meals a day in the mess hall. I carpentered until we got the carpenter work done, electrical work until that was done. Then I put up the conveyor belt by myself. The veins had all been discovered before this last time, and they had worked them out. They went under the veins, shot and hauled them out.

When it got up higher than you could reach, they timbered up; they would get up in there and drill out holes about ten feet deep, and shoot that ore down on them timbers. They had loading chutes. They would open the gates and let the boxcars fill. They would close the gate and take it off, go back and get another. They ran a pretty regular load. The mill ran seven days a week, twenty-four hours a day. We would shut down about one day a month to clean up, to get the sediment ore out of the machines. About once a year, we would have to reline the ball mills. After the wage an hour law came in, we got $2.40 for eight hours. Before that we worked for $1.50 for twelve hours.

Henry Wallace Green

I went to work in April 1935 in the assay office, working up samples that came out of ore from the ball mill and get the average of what the value per ton of ore [was]. The process consisted of crushing rocks, pulverizing and heating [them] to get the moisture out. We mixed it with a concentrate and put it in a 1,200-degree furnace and burn[ed] it down. Then we put it in a cupula and taken the lead out of it. We had to put silver with it to catch the gold. Then, we'd take and put the silver and gold in acid to take the silver out, which left the pure gold. It was weighed on a scale that would weigh anything that had any weight to it—probably a hundredth of an ounce.

We crushed the samples in one building and worked them up in another one on account of so much dust and stuff. But the scales had to be on a post that was in the ground that came up in the building. The building couldn't even touch 'em because any vibration, would knock the scales off. They were that sensitive. It was real interesting.

In the first building was where we had to crush our rocks. We'd crush them like marbles and then dry 'em at a hot plate. Then we'd run that through a pulverizer and grind the rocks into a powder. Then we had to dry it again. Then we'd take a portion of that and put it in a little envelope with

Alabama Gold

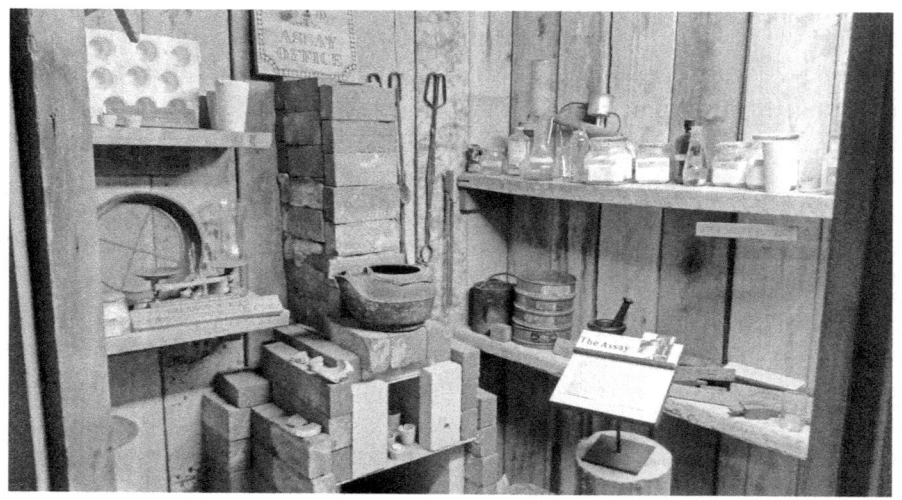

This picture of a model assay office is used with permission of Pine Mountain Gold Museum, located in Villa Rica, Georgia. *Courtesy of Peggy Jackson Walls.*

a vein number on each one. What I mean is it went by letters like SEA or something like that. But they went by letters. So the ball mill samples and the concentrate went by letters. We didn't use all that for a sample, but we kept it in case that sample messed up or [if] we had run another one, we had enough stuff to run it. And that's how they got the average of what there was per ton in the ore. The first of each month they would take what they called face samples. If they hit a new vein they'd check it to see if it was worth drilling. They could get the average by taking a sample of it, the vein, and run it through.

I made fifty-two dollars a month working seven days a week. We didn't have certain hours in the assay office because there wasn't but three of us working there, but the hours were set up seven to three. But sometimes we'd get done by twelve o'clock. So when we got done, we were off. Sometimes we had to work 'til four or five that evening. It was set up for seven days. If the mine just run five, we had two days we'd just work a half a day. We'd work seven days a week.

The ball mill run seven days a week unless it run out of ore. But most of the time they had a big ore bend. Held maybe a hundred tons a head. The ore was ground into fine consistency and shipped in cans to a train in Alexander City.

At the time, we were living on Campground Road at Jackson's Cross Roads, about five miles from the mountain. We walked part of the time and

Surviving the Depression

finally got enough money up to buy an ole '28 Chevrolet from my brother Llewelyn. We went to work in it until the bridge fell in. They got the bridge fixed where you could walk across it, but it was out for a long time. We'd get on what they called a skeeter and go from there on to the mine. We walked when the ground was spewed up and ice all over everything. My brother Carl and I worked together in the assay office, and Llewellyn worked at the mine. He worked different shifts 'cause they run it straight through twenty-four hours where he worked at the flotation machine.

Oplin was my boss man. He was from South Carolina, and he worked with us until we understood what to do in the assay office. When the mine closed, Thomas Daughtery went on to the Carolinas to work at a gold mine there. Carl went to a gold mine between Talladega and Ashland.

The company had to sign an agreement with Alabama Power, agreeing to use their power for five years before they would build a line. The mining company had to furnish so much help to clean the right of way out to put the lineup. The mining company had to put up a deposit before Alabama Power would put up the lines. The lines went just to the mines. There were no offspring lines to houses in the community.

The rolling store would stop at everybody's house. If you had chicken or eggs, you could swap them for groceries, some merchandise, a few clothes. We didn't have a lot of money to buy clothes then. I remember Mr. Eason and the others. Mama kept boarders [who] worked at the mine, usually, four at a time. And that's where she'd buy her food supply—off the rolling store. Rob Bowen kept some boarders. Miss Daisy Simpson kept boarders, too.

8
Life in a Gold Mining Community

Five years after Mr. Johnson discovered gold at Hog Mountain, thousands of men were digging in the hills in northeast Tallapoosa County. Goldville alone had an estimated 3,000 to 3,500 inhabitants. A number of families grew tired of the boisterous environment with saloons and race tracks and left in search of a new site. They found one a few miles south, where they set down stakes and built a town with stores, sawmills and cotton gins. By 1857, the town was officially known as New Site and home to a US Post Office. On the west side of Hog Mountain, a little farther removed from the rowdy camp life around the gold mines, pioneer families settled on farms and built community institutions of churches, schools and stores. In the Hackneyville area, the first structure was a justice of the peace office with Sam Belle presiding. Barney Kemp was constable. A saloon was built nearby with a Mr. Pennington serving as proprietor. All the buildings were built from hand-hewn logs and roughly put together. Jim Nelson owned the property at the foot of Hog Mountain near Hillabee Creek, where he maintained supplies for the mining camps nearby. He also reserved one room in his home to serve as the community store.

Several churches were built in the communities of Hackneyville, Cowpens, Valley Grove, Goldville and New Site. Near Hog Mountain, families attended the Hillabee Campground Church that took the name Hillabee from the creek around which the Creek Indian culture flourished for hundreds of years. Church services were held under a brush arbor as early as 1847, when gold mining activity was widespread in northeast

Alabama Gold

Tallapoosa County. Pioneer families arrived on horseback or in buggies and wagons for camp meetings after crops were harvested, and tomatoes, peas, beans and a medley of other vegetables were canned. Camp meetings were held under the church arbor until the first frame structure was built in 1852. Adults sat on split log benches; and for additional seating, children sat on the straw-covered ground. Candles made from beef tallow lit the gathering of friends and neighbors in the evenings. For the fall camp meetings, temporary bunkhouses were constructed near the natural spring and near the arbor to supply water for cooking and bathing. One building was for the men and boys and a second for the ladies and girls. They sang familiar songs, such as "The Old Rugged Cross," and listened to several sermons throughout the day and evening. They also found ample time to visit and catch up on community news. When the week was over, the people loaded their belongings into their wagons and surreys and headed home, refreshed. The late summer meetings continued well into the twentieth century until fear of gypsies and horse traders passing through the area frightened residents. Although the fall campground meetings ended, the old campground church still stands today with an active membership. Weathered headstones in the small cemetery date back to the early 1800s. One of the earliest is Wild Bill

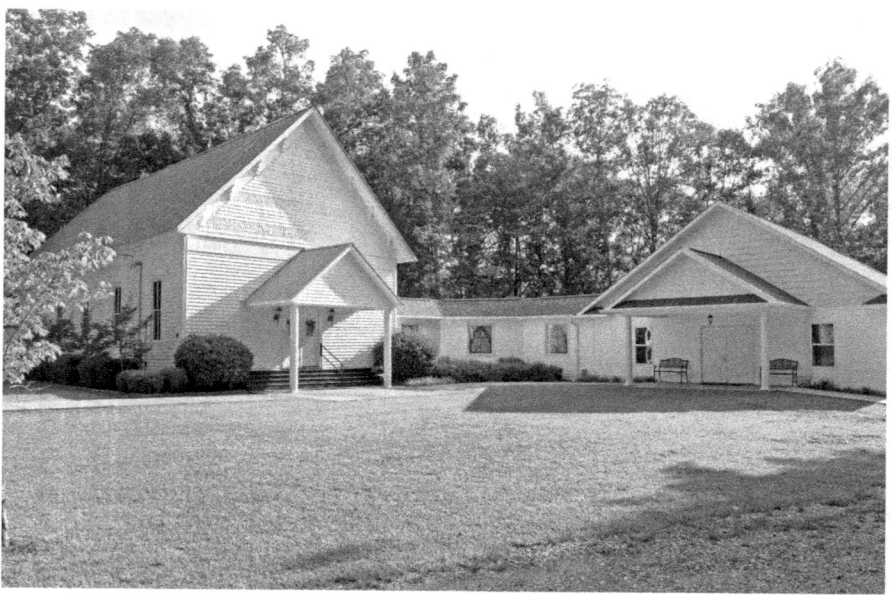

Hillabee Campground First United Methodist Church dates back to settlement days when only an arbor was used for meetings. *Courtesy of Peggy Jackson Walls.*

Surviving the Depression

Hutchinson, who lived with the Hillabee Indians and panned for gold in the Hillabee Creek.

In Hackneyville, Presbyterians, Baptists and Methodists all went to the same church and took turns using denominational materials until they were able to build a Baptist church and a Methodist church in the early twentieth century.

The Hershel Baker Store

Lynwood Baker and her husband, Dan, operated the Baker store, where they got a lot of business from the gold miners. Lynwood's mother, Daisy Simpson, operated a boarding home for the miners:

> *I remember in '32, '33, '34, seeing the truck go by with the big cans on it. The truck went by at least twice a day. There would be two layers of*

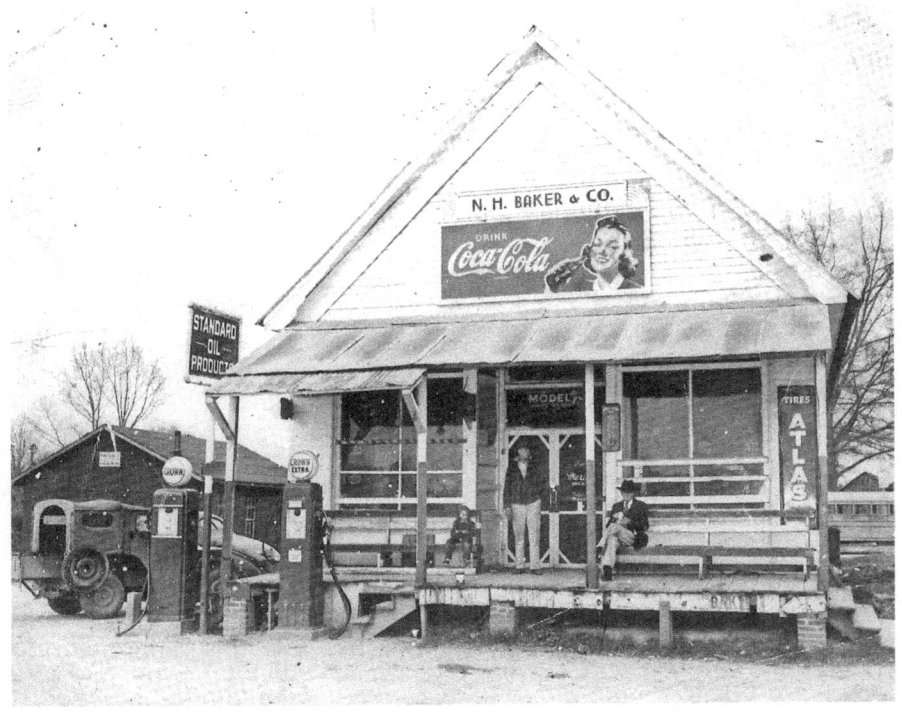

On the left are Betty Baker and Dan Baker (standing); on the right, Neil H. Baker sits at the Baker country store in Downtown Hackneyville in the mid-1940s. *Courtesy of Betty Baker Hamilton.*

cans on the back of the truck. Mama [Daisy Simpson] kept boarders, and sometimes they would stop by. We would go see the stuff in the cans. It looked like dark powder. These are my memories when Mama kept boarders. The mine was two to two and one-half miles from our house. My brother quit school and went to work at the gold mine when he was sixteen.

Alvin Goodwin drove the truck, and this big German shepherd dog rode with him. When the truck passed by, he would be sitting up in the truck. The dog carried messages for them at the mine. It was real smart. They'd put a note on his collar, and they'd tell him where to go. There were a lot of buildings over there. I think Alvin was the one driving when the Sanford Bridge fell in, and the passenger was killed.

Miss Daisy Simpson

Miss Daisy kept boarders when the gold mine opened in 1933. Along with taking care of her husband, Ernest Lynwood Simpson, and their three children, she stayed busy with the farm and household responsibilities. Vegetable gardens had to be weeded and cared for, particularly in dry weather, when water was carried from the well or spring to keep the plants from dying. Trips to the garden for fresh peas, okra, corn and tomatoes in season were daily chores. Peas had to be shelled, corn shucked and okra and tomatoes sliced in preparation for a meal. Daily trips were made to the barn to milk the cow and to the henhouse to retrieve eggs for breakfast and to use in cooking. A great deal of work went into preparing meals. Miss Daisy described her daily routine, which included cooking for boarders who worked on different shifts:

> *Lots of nights I'd get up at eleven or twelve o'clock and cook them something to eat. They'd work their shift and sleep. I'd have to have something for them to eat about two o'clock to go on their job. And it was work all the time.*
>
> *Life was hard along then. Everybody had to work to live.*

Surviving the Depression

From One-Room Schools to Public Schools

Former governor John M. Patterson, a Goldville native, described the education available in the backwoods of Alabama in the late 1800s and early 1900s. Judge Patterson is the son of Albert Patterson, best remembered for his attempt to clean up Phenix City's corruption and his assassination by organized crime, an act that catapulted John M. Patterson into state politics. He served as Alabama attorney general (1955–59), governor (1959–63) and federal judge (1984–97). In 2003, he served as chief justice of the special Supreme Court case of then Chief Justice Roy Moore, who was forced to remove the Ten Commandments monument he had placed in the Supreme Court building.

Governor Patterson's long and distinguished career in public service had its roots in humble beginnings in northeast Tallapoosa County. His great-grandfather John Love Patterson settled near Hillabee Campground, where he operated a gristmill. The governor's grandfather Delona Patterson, like other people of his day, had little opportunity for an education although they strongly desired one.

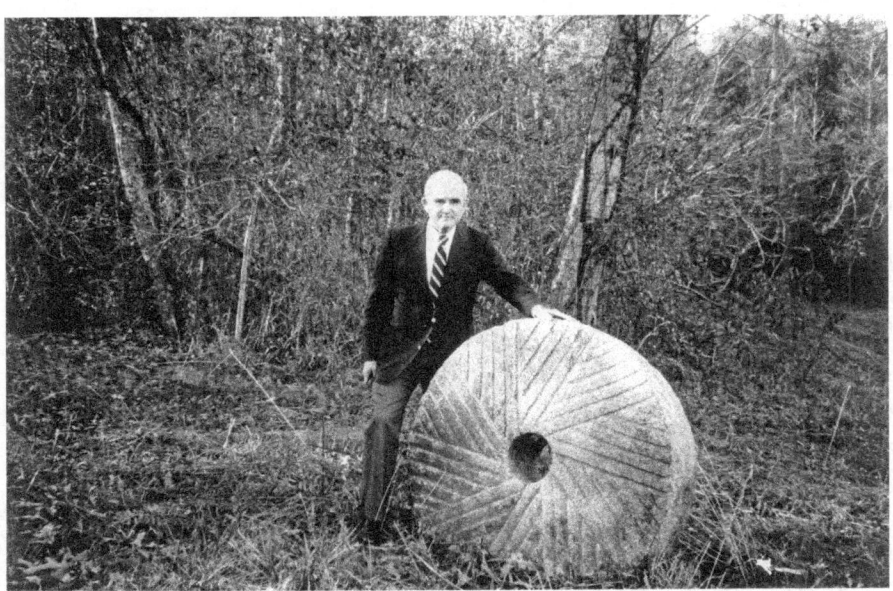

John Malcolm Patterson (governor of Alabama from 1959 to 1963) beside the millstone used by his great-grandfather John Love Patterson, a miller and early settler of Tallapoosa County. *Courtesy of Peggy Jackson Walls.*

Alabama Gold

People seemed to have a thirst for knowledge in those days, but a formal education was hardly available since there weren't any public schools. Grandpa Patterson never went to school but just a few days in his whole life…through teaching himself, he became a pretty well-educated man. He and others in the community built a two-room schoolhouse at Goldville. Another was built at Valley Grove and one near Hillabee Creek [Hot Chapel]. There were no public schools at the time. People in the community chipped in and paid the teacher's salary. The teacher boarded at one of the families' homes until the end of the school term, which was short. All of the grades were taught in one or two rooms, basics like reading, writing, arithmetic, and spelling. My father [Albert Patterson] became a teacher that way and would board in different places. My mother did the same thing and my uncle Lafayette Patterson.

Miss Alice Foshee

"Miss Alice" attended Hot Chapel school, which was within walking distance of her home. At that time students had to live within walking distance of a school or board with someone who did. Her memories of attending

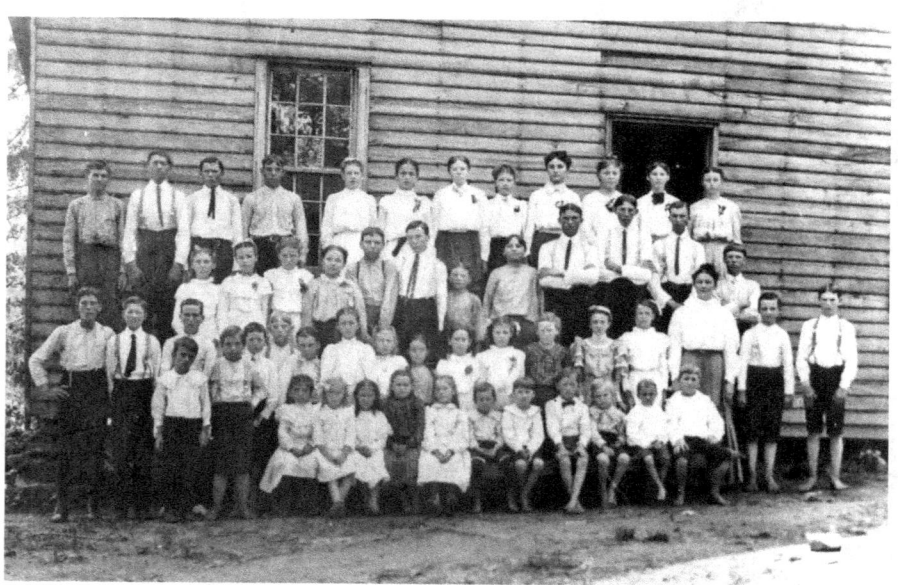

Students of all ages pose at the Hot Springs one-room school near Hillabee Creek on the old Nelson place. *Courtesy of David and Karen Daniel.*

Surviving the Depression

school present a picture of a typical one-to-three room building, similar to a sharecropper's house.

Hot Chapel was located about one-fourth mile southwest of the old covered Sanford Bridge. The building was made of weatherboard pine. There were three rooms. The large room had a fireplace on one end and a heater in the other. [There was] *pitiful heating though. People* [who] *had kids in school furnished the wood. Half the time, they were out. For a blackboard, the teacher used a large square of the wall that had been scrubbed smooth and painted black. Students brought lunch in aluminum buckets and hung them on a nail on the wall. The desks were so constructed that two students could sit in each desk. Hot Chapel* [went] *from the first grade through the ninth. The building consisted of one large room, divided into two classrooms by a curtain in the center. The boys and girls entered through separate doors to their side of the room. In the winter, the boys' side was heated by a wood heater, and the girls' side was heated by a large open fireplace. Students were charged a fee of $2.50 each month. Children who lived close enough would walk home for lunch and then walk back to school. At recess, students played paddlecat or soft ball. The older girls did the batting, and the smaller girls would run for them. The bathroom facilities were located far down the hill from the school.*

Mr. C.E. Newman

When I came to Hackneyville in 1926, the school was an old two-story building. There were no electric lights, and the building was heated by old wood stoves. Parents brought in wood by the wagonload to keep the school warm for the children...Since I was at Hackneyville during the Depression years, I remember the highest salary I ever got was $150 a month or $1,800 a year. But during the Depression, I didn't get a payday for eight months. We were paid in scrip. Merchants would give us fifty cents on the dollar for the scrip money, but they wouldn't cash it. For the first few months, merchants in the community took teachers on credit and charged them interest, but they didn't have money to buy items to sell if folks didn't pay them. I remember some of our high school children dropped out of school and got a job making twice as much as I did teaching.

The high school students would learn and give a different play every two weeks [as a fundraiser for the school]. *Admission ran ten to fifteen cents, unless it was a royalty play; then it might be twenty-five or fifty cents. All the money went directly to the school. If it was a hit, we would take it to other schools in Alex City and Rockford. We had basketball games in Alex City in Russell auditorium and charged admission.*

Company Doctor

The Hog Mountain Mining and Milling Company doctor was Dr. T.H. Street, but it was his assistant, Dr. James E. Cameron, who made most of the calls to the mine. After completing his medical training in 1935, Dr. Cameron worked closely with Dr. T.H. Street, Dr. J.J. Walls and Dr. Wade Lamberth and learned quickly how to treat patients without the benefit of antibiotics or other modern medicines.

Between August 1935 and January 1936, Dr. Cameron made many trips to the mine. Occasionally, it was necessary for him to go down into the mine to examine and treat a miner. Dr. Cameron became good friends with Superintendents Neal Johnson and Elmer Alderfer, both of whom were graduates of the Colorado School of Mines. Often they came into town on the weekend and met with Dr. Cameron and John Coley at the Russell Hotel and played bridge.

Dr. Cameron reflected on his experience as Dr. Street's assistant, as well as Dr. Street's appearance and personality traits:

> *Dr. Street was a very large man and a very particular dresser. He bathed and changed clothes twice a day.*
>
> *If it was raining and wet and Dr. Street thought he might get his shoes dirty around the mine, he would ask me to go up to the mine so he wouldn't get his clothes dirty. Dr. Street drove a Dodge automobile and bought a new car every year from Wilbanks Motor Company in Alexander City.*
>
> *I made house calls day and night up there in that area, sometimes when the roads were nearly impassable, and they often got very bad in the winter. The people who lived up there were the "salt of the earth." They were good, solid people, but they had such a little bit of money with which to pay for anything, not just a doctor, and the care we gave them wasn't worth a great deal. They would have gotten well just about as well without the medical care they got.*
>
> *I remember one miner who had his foot caught when some timbers fell down in the mine. I went into the mine and gave him some morphine to*

Surviving the Depression

alleviate his pain before they pried off the timbers to get his foot loose. The managers were afraid he might lose his foot, so they were very solicitous of his well-being.

Eugene Leslie

A number of mining accidents occurred at Hog Mountain in the 1930s. Eugene Leslie, a timberman, shared his experience in a serious accident over two hundred feet down in the mine.

My name is Eugene Leslie. I was born in 1898. I worked at Hog Mountain gold mine for several years and was underground when a fellow took an axe and hit a rock and it hit me in the eye. They sent me to St. Margaret's hospital in Montgomery. I stayed there a long time. That like to have killed me. They gave me a shot and took the eye out. All I could get was compensation. It wasn't much. You know, we used to couldn't get nothing.

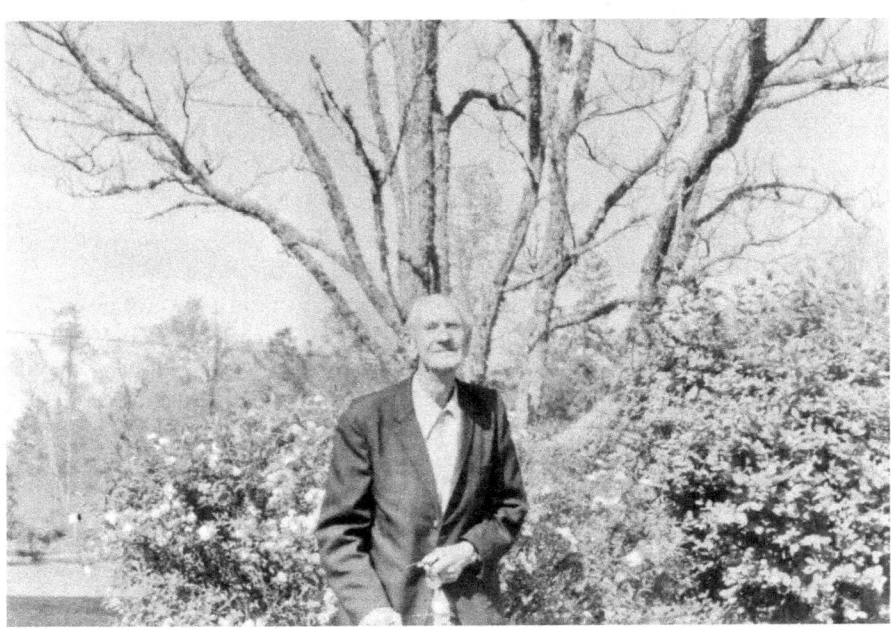

Eugene Leslie was a Hog Mountain miner in the 1930s. He lost an eye in mining accident at Hog Mountain in that decade. *Courtesy of Peggy Jackson Walls.*

Alabama Gold

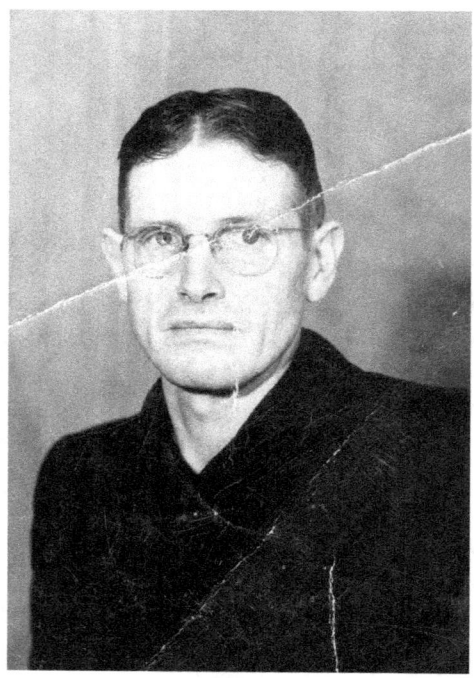

Hog Mountain miner Kermit Jackson was head timberman and driller and brother to Benjamin Talmadge Jackson. *Courtesy of Peggy Jackson Walls.*

I timbered with Kermit Jackson and old Buck Patterson. They was so stout, they would say, "Hold it a minute, let us get this." They sure were good workers. Kermit and Buck would cut big hickory trees on the mountain. They'd put the timber on the elevator themselves. That elevator went down 200 to 225 feet. Then sometimes, I worked with Marion Bowens to build stopes to catch the rock they'd shoot down. We'd set charges off every day. They'd shoot for thirty minutes, one right after another you could hear them. They sounded like they'd shake that mountain down. Didn't nobody stay in there when they lit them; they got back out of the way. After the charges were set off, in the evening, you wouldn't go back I 'til morning. You had to have the alphabet to know where you were going: "A right A, A right E, and A right B. A left E, A left A, and A left B." The veins came out of the main stoke.

Mr. Brown, the general superintendent, said at the time the mine was shutting down there was money set aside there, and I ought to have got more for the loss of my eye, but I would have had to sue them to get it. If I had, I couldn't get a job with a company anywhere. I had to make me a job, so I started painting. But when Mr. Brown come back, he came on down and hired me to be a rigger.

Mr. Brown was a fine man. The best man I ever worked for. He was an electrical engineer, a gold mine graduate from New York. The government gave him his authority. Whatever he said went. There's gold there now. I'm gonna tell you the reason they shut it down. You'd have get rid of this waste rock to get to the gold, and it costs money to do that. The stockholders made money, and they'd hold the money. That's what shut it down. They wouldn't take any more chances. I had more confidence about what was in there than

Surviving the Depression

anybody. I got that straight from George. "Someday Hog Mountain will open up," he said. "There's still gold down there. It'd run $37 a ton." I also helped at the graphite plant in Millerville. That was during the war [World War II]. We had to have graphite to crucify steel. After the graphite plant, we went north to Henderson, North Carolina, at Thompson's Steel, to build a plant to get tin, gold and zinc out. When we got started, General Electric bought them out. When I come home, and Mr. Brown come back down here, he wanted me to go to Colorado with him and work at the mines up there. But I stayed home this time to be near my family.

The Buck Patterson Story

Buck was in a fight somewhere, and the sheriff come to get him. He run and jumped in the elevator and went down in the mine. Well, they just kept it going 'til there wasn't no way to get Buck out. Kermit talked to him; I did, too. Told him, "Buck, they'll starve you out. They'll stay right there and can't nobody bring you food. The best thing to do is give up and go on." They arrested him, but he finally got out of whatever it was. He come back over there and worked awhile, then left after that and was deputy sheriff somewhere for a while. He got in trouble again, and when the law come to get him, this time he killed three men. They give him a lifetime sentence for killing them.

9
Notable People and Events

Although many residents in the Goldville/Hog Mountain area have interesting stories to recount about their lives and those of family members and neighbors, a few are well known in Alabama's politics, such as Governor Patterson; his uncle Lafayette; and his father, Albert Patterson. Judge John Patterson fits both categories. He grew up on Patterson property that was purchased in the 1800s by his grandfather.

> *Around Goldville...was what was known as the Goldville Field. The mining commenced here about 1840, and by 1843, there was quite a large settlement at Goldville. In fact, there was over three thousand people there at that time. The post office there did more business than the post office in Montgomery. There were, I think, twelve stores, a hotel, a sort of a tent city, and a race track. There was quite a lot of gold mining activity going on around here at that time. There were no records on that. In those days, there were no tax laws, and no regulations. People traded in gold bullion. Really, you don't know what they found here. Only the experts can estimate as to what they might have found here.*
>
> *Now I never saw any mining here myself. The only mining I ever observed here was when I was in high school in Phenix City. I'd come up here and spend the summers with my grandparents in Goldville. In 1935, '36 or as early as 1934, there was a mining operation going on here where they were hauling the old tailings from the mine over at Hog Mountain. The tailings were the ore that had been originally worked, and there was a*

lot of stuff left in it. They were hauling it over here at Goldville and putting it in concrete vats and reworking it with a new process. Apparently, they found that to be profitable. I went over there one day with my uncle James Harry in a wagon to see it. I was fascinated by what was going on, and that's the only actual mining operation I ever saw here at Goldville.

There is an area here at Goldville, on my place, called the Houston Mines. And these are mining shafts that go straight down, then tunneling out following the ore, and you can still observe these. Some of them are very deep and almost out of sight. Some of them are slanted. Some of them are approximately twelve feet square. It is my understanding that some Welsh miners came in here after the initial mining. It was their technique to build these twelve-foot-square shafts. You can go and observe these and see that there was an extensive mining operation. You can hardly go a hundred yds. around Goldville without finding some diggings where people were trying to find gold, apparently working claims. Right near my gate where you come in here, if you step off the road about twenty feet and start walking up in the woods, you'll see extensive diggings, which reflects that a great deal was going on here. It is my understanding though that they struck gold in California along about 1849, and most of the people here that were interested in gold mining packed up and moved west, where it was a richer strike. And Goldville began to fade away as a city.

My uncle Lafayette Patterson, at ninety-nine years old remembered Goldville as a thriving place with businesses. He remembered a circus coming to Goldville and setting up in the streets of Goldville with performers and animals. Goldville had a great day at one time, but I think today there's something like fifty people now, counting the children.

My great-great-grandfather John Patterson and his brother Malcolm Patterson moved to Alabama about 1831. Malcolm settled with his family in the Hatchett Creek area, just north of Goodwater. John Patterson, whose full name was John Graham Patterson settled along Town Creek near Hackneyville with his wife, Mary Love. They had five children. My great-grandfather John Love Patterson inherited the farm and built a gristmill on Town Creek. John Love Patterson was a miller, and he was exempted in the early part of the Civil War. [Of] course, he was getting on up in years. I don't believe that he ever got any farther away than a big camp at Lochepocka. When the Federal troops came through, they did take him and the other soldiers over to Phenix City, and he did fight in that last battle at Jerrod, such as it was. I don't think there was much of a fight.

Surviving the Depression

[Great-]*Grandpa Love Patterson was five years old when the Union troops came through Alex City, and he was in town that day with some of the black slaves in a wagon picking up supplies for the farm. He told that they got afraid the soldiers would come and take their wagon and mules. So the black fellow drove the wagon and mule as hard as he could to get home and off the road before the soldiers came by.*

I just can't imagine how they must have lived in those days. They must have had a section of land and operated a gristmill and had four or five slaves. They were bound to be fairly well off for somebody to be up here in this pineywoods country. [Of] course the farm was self-sustaining in those days, a big one, and it was a pretty good size—six hundred acres. Cotton was the cash crop. They grew corn to feed the livestock.

My great-grandfather John Love Patterson and a lady named Ellen Fields had a child named Delona Patterson. [Ellen] either died or left with her family and went to Texas, leaving young Delona Patterson with his father, John Love Patterson, to raise.

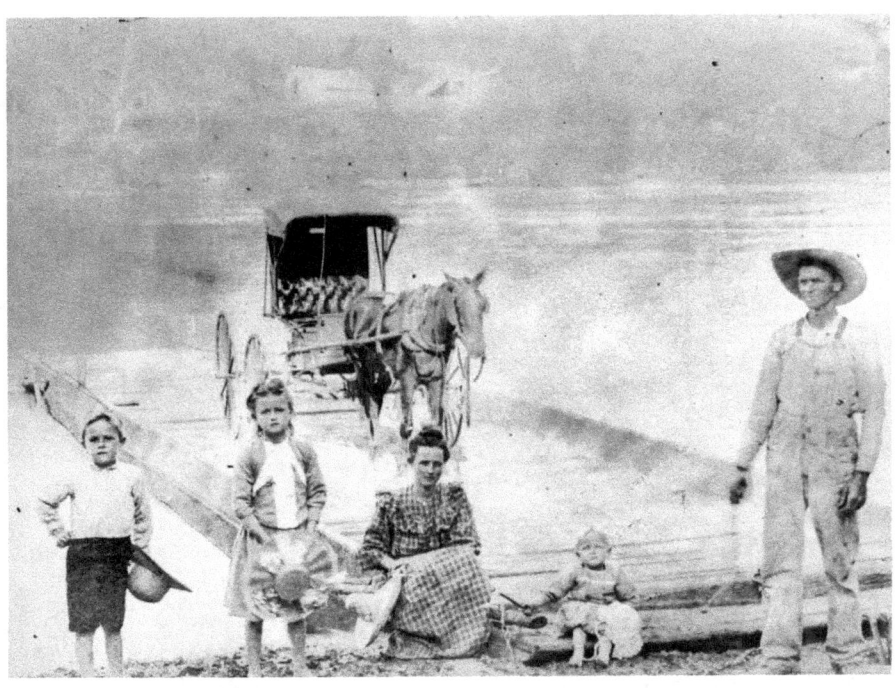

Miller's Ferry is an example of how goods and families were transported across the river when few bridges were built. *Courtesy of Tallapoosee County Historical Museum.*

Alabama Gold

I can remember when there were no paved roads at all, no paved roads leading out of Alexander City in any direction. If it rained two or three days straight, you couldn't [sic] hardly get to Dadeville and to go to Birmingham would take all day. And you could either cross the bridge at Childersburg and pay fifty cents toll, or you could go down the river and take a ferry for a quarter…My father used to go take the ferry to save a quarter. That's how valuable twenty-five cents was then. But up in here was really back in the country those days. I was born here on this place and stayed here 'til I was old enough to go to school. My parents were schoolteachers. I would go and stay with them during the school year and come back here as soon as the school year was over.

Lafayette Patterson

Lafayette remembered where the old racetrack and the old hotel were, as well as the general layout of the small town Goldville, once home to thousands of gold miners:

The town was mostly farmed by dirt diggers. The population that once made Goldville one of the largest towns in Alabama dwindled to almost nothing. Hog Mountain, Low McGregor, [Log Pit], Birdson Creek Mines once dotted the red hills around Goldville, with mine tailings piled high outside the entrance of many of those mines. The mines worked night and day by light of crude lamps until news reached central Alabama in 1849 of the California strike.

Judge C.J. Coley

A native of Alexander City, Judge Coley was involved in preserving Alabama history and served on the Alabama Department of Archives and History board of trustees for forty years and as its chairman in 1996. He was instrumental in creating the Horseshoe Bend National Military Park and wrote several articles for historical journals, such as the *Alabama Review* and the *Alabama Historical Quarterly*. Fascinated with the gold mining history of Alabama, he wrote a paper titled "The Climax of Gold Mining

Surviving the Depression

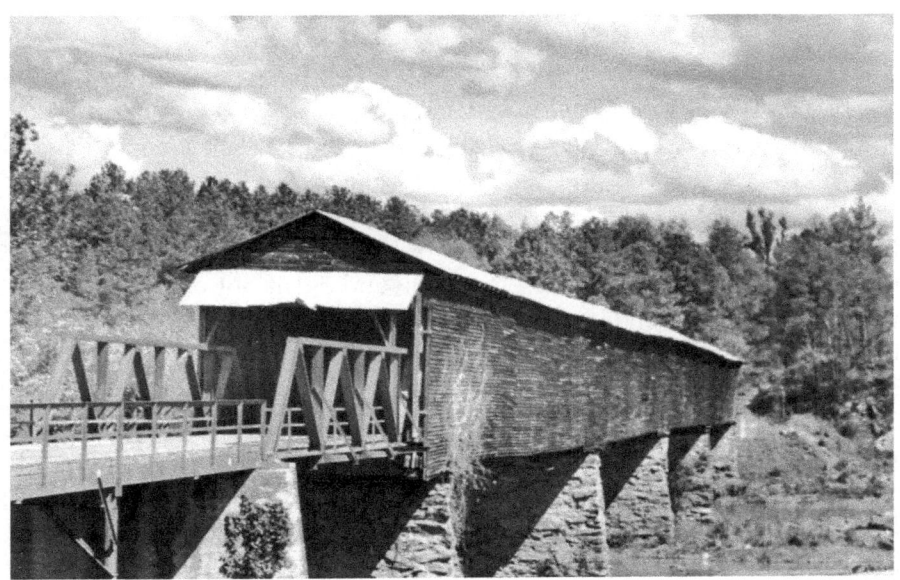

The Horseshoe Bend Covered Bridge was built with wooden pegs and lumber near the Horseshoe Bend battle field and was destroyed in the 1990s. *Courtesy of Tallapoosee County Historical Museum.*

in Alabama" and presented the paper to the Alabama Historical Society on May 5, 1967, in Mobile, Alabama. He noted the fluctuations in the gold mining industry and concluded that more money might have been made from the sale of land than from the discovery of gold, observing, "In so many cases men who got bitten by the gold bug would be able somehow to raise considerable money to open and operate a mine, and when the operation proved unprofitable a sign was hung 'Land for Sale.'" To elevate the price of his property and to entice a prospective buyer, the landowner would shoot gold into the ore by removing the shot out of a shotgun shell and filling it with gold particles, which he then fired into the crevices of rocks, where the gold particles could be seen. Using another method, called salting, the seller scattered gold in areas that could easily be seen by anyone examining the property. Excitement drove the gold mining industry and attracted those who were looking for a quick path to wealth.[74]

Alabama Gold

Cowboy Osborn

Cowboy Osborn worked for the Hillabee Gold Mining Company before it closed in 1916. His personality and his story could have been a page out of Johnson Jones Hooper's Simon Suggs, Captain of the Tallapoosee Volunteers.

My full name is James Orland Gaines Osborn. There used to be a fellow named Bill Arp writing a story in Home and Farm out of Georgia. He was writing a story about a fellow named Orland Hyde. So when I was born, Papa named me after Orland Hyde. I was born [on] September 27, 1888. ... When I was in school, Woodson Galloway and I were big friends. Woodson went to Hog Mountain, working in the mine. He got to writing to me, wanting me to come up there and work with him. Well, I quit school, come to Hog Mountain and went to work for Hillabee Gold Mining Company. We made a dollar and ten cents a day. Nobody couldn't get that nowhere else. We were just teenage boys. Them boss men let us work together in the mines there, shooting dynamite. We never had no experience or nuthin'. We like to a got killed once. Never forgot that, but

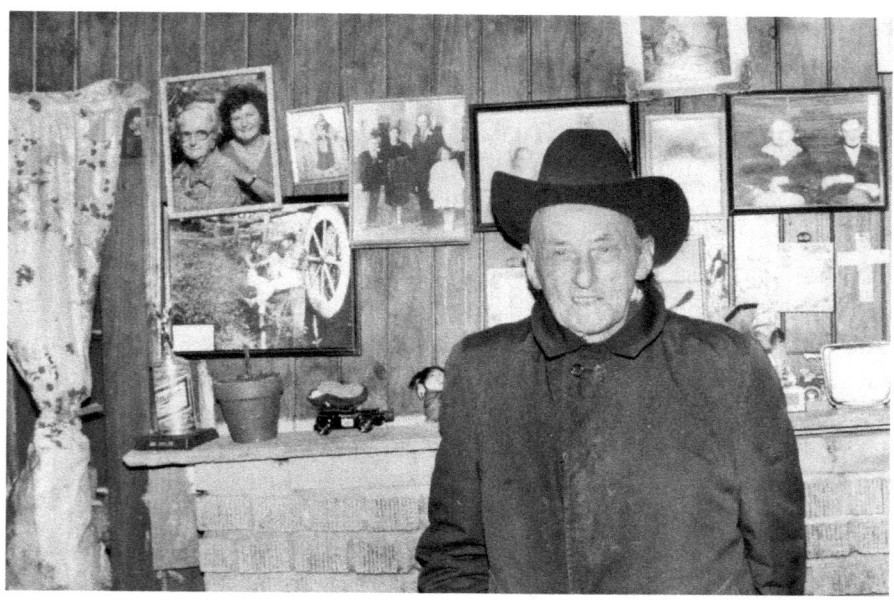

James Orland Hyde "Cowboy" Osborn was a Hog Mountain miner who worked in the pre–World War I operation as a teenager with his friend Woodson Galloway. *Courtesy of Peggy Jackson Walls.*

Surviving the Depression

they just let us shoot that dynamite. That'd be agin' the law now. We 'us 'bout sixteen or seventeen.

They'd dig a big cut deep without no top over it [and] then maybe start to tunnel under. That's the way was; we had to go way under there. Anyhow, we had to follow the vein, you know, if it went straight up, you had to just keep working out. If you weren't careful, you went out the top of the mountain. And we like to have went out of the top of the mountain. We got under a big, longleaf pine tree. Got to the taproots of it. We was doing our best; we didn't want them to know it, but we thought we'd keep on and have that pine fall down in the mine. The old boss man came up and saw we worked out the top. He separated us then, put Gallaway with someone else. When they changed me, they put me with an old man named Tucker Cash, seventy-five years old. I never heard no cussing much or nothing. They put me there with him, and he didn't have no mercy on me with me just being a kid and all. We used old wooden wheelbarrows, and he'd load that thing like I was a big man, you know. He'd just pile it on there, and I just barely could move it.

And so when they'd shoot, the boss man would take his pick and pick overhead to see if there was anything loose to keep it from falling on us. So I was standing out there a little piece from him one day and he kept picking and picking and struck a loose rock. He kept on until that thing came down. It just came down by his legs, scrapped them. He couldn't walk hardly; he was already so crippled up so much had already fell on him. I never heard so much cussing in all my life. It like to have scared me to death.

I'd shoot; then in a little bit, we'd go back in there and go back to work. That old dynamite smoke like to have killed me. They finally had to carry me home. Felt like it just lifted up my head off me. So I didn't go back no more. When I got to where I could, I started back to school again.

Well, see we boarded out there. We paid ten dollars a month. It was just like a town. Folks lived all over the place with houses and all, a company store and little bitty houses all around it. They had an old commissary, the highest-priced place I ever saw in my life. We'd come out of the mines at twelve o'clock and sit out there. They had a Negro cook, and she'd bring cabbage and big ole white biscuits and no cornbread. I just learned to eat what she brought us. My main boss was Mr. Hill Watson, and the other was Mr. Smith. Walter Neal Barfield was the old Negro who drove the wagon that hauled the ore. But this last time [Depression era] the mine run, they had big trucks and folks hauled ore in the trucks. Jim Nelson'd drive the gold mine wagon, [as] they called it. He done the hauling from the mine to the Alex City depot, and they shipped it somewhere. And one time, I heard somebody say he had twenty pounds of gold on that wagon. It'd be dangerous now hauling that much gold.

Hog Mountain's run three times that I know of. Well this last time, I didn't work. The first time, Papa had a large garden, and he put out a lot of strawberries. That's where he first started truck farming. When the strawberries got ripe, he got to picking large buckets full of strawberries, and he'd tote 'em clear to Hog Mountain and peddle 'em out. They were glad to get them. That was his start of truck farming.

Cowboy was friends with US legislator Lafayette Patterson, Governor John Patterson, attorney general–elect Albert Patterson and state senator John Harlan, all from around the Goldville and Hog Mountain area. He missed the 100th birthday party celebration Governor Patterson promised him by only a few months. Cowboy's many friends, his sense of humor and the occasional glass of homemade wine kept him content. When he died, he was a beloved neighbor in the Valley Grove community and as memorable a character as the original Orland Hyde whom Bill Arp wrote about in the late 1800s.

Ben Russell

Ben Russell is the son of Robert C. Russell, owner of Dutch Bend Mines and grandson of Benjamin Russell, known as "Mr. Ben," founder of Russell Manufacturing Company, and Roberta Bacon McDonald "Miss Rob."

"Miss Rob" was the first in the Russell family to become interested in gold mining. She influenced my father, Robert Russell, and my grandfather Benjamin Russell to invest in the gold mines out West. I remember hearing my father and McKinley Hoyt talk about Dutch Bend. McKinley managed the mine, hired people to work and checked on Saturdays and Sundays to make sure the water pumps were working. This was a concern because there were multiple levels below the water table, where they had drilled down in the ground, branching off into tunnels. The site still contains the large concrete and rock foundation for large machinery [that] was there long before my time.

I heard stories about Dr. Ulrich, who bought some land in northeast Tallapoosa County he thought would be suitable for growing grapes and making wine. When he and his party dug into the hillside, they discovered gold and began gold mining. Ulrich made gold bars he exchanged for goods.

Surviving the Depression

He did not make shipments to a mint, nor did he altogether abandon his dream of building a commercial wine business. He had a grape vineyard and made wine, although any records on this business were lost. Dr. Ulrich was a German immigrant, but local people thought he was Dutch. This is why they called the bend in Hillabee Creek and the mine itself Dutch Bend.

My father had a lab in the basement of our home with bottles of chemicals, including mercury. With the help of his good friends Dr. Walter B. Jones, state geologist, and Charles Dean, a mining expert, he assayed ore samples. He had a small pipe running out of the basement to remove the chemical fumes and cautioned me, "Mercury will kill you."

After the Dutch Bend mine closed, my father operated a mine nearby, about 300 feet SW, across from the Rocky Creek Baptist church. There was a building with four levels where on each, a different step in processing ore was completed. Approximately six to eight people worked at the mine. McKinley Hoyt managed this mine for my father.

The Dutch Bend property is owned by Russell Lands Inc. The Russell family has never owned any part of Hog Mountain; however, Benjamin Russell was on the board of directors for the Hog Mountain Mining and Milling Company in the 1930s, and gold miners cashed their checks at "Mr. Russell's bank in town," later known as First National Bank.

The Mysterious Caretaker

Mrs. Marie Aldrich Cravener was the daughter of Colonel T.H. Aldrich Sr., who, with his son, T.H. Aldrich Jr., operated the Hog Mountain mine from the early 1890s through 1916. Ownership of the mine was transferred into a corporation known as the Hillabee Gold Mining Company and operated under this name from 1905 until 1916. T.H. Aldrich was an engineer who continued to work in the field of mining and sold his share of the mine to his sister, Marie, making her the sole owner. After the Hog Mountain Mining and Milling Company closed in 1937, Marie began to search for a caretaker who would live on the mountain and look after the mine and the timber. For about two years prior to Mr. Neil's becoming the caretaker of the Hog Mountain property, he would come to Hog Mountain and "look around." When the Marie asked people of the area about someone who would be interested in taking care of the Hog Mountain property, they recommended Mr. Neil.

Alabama Gold

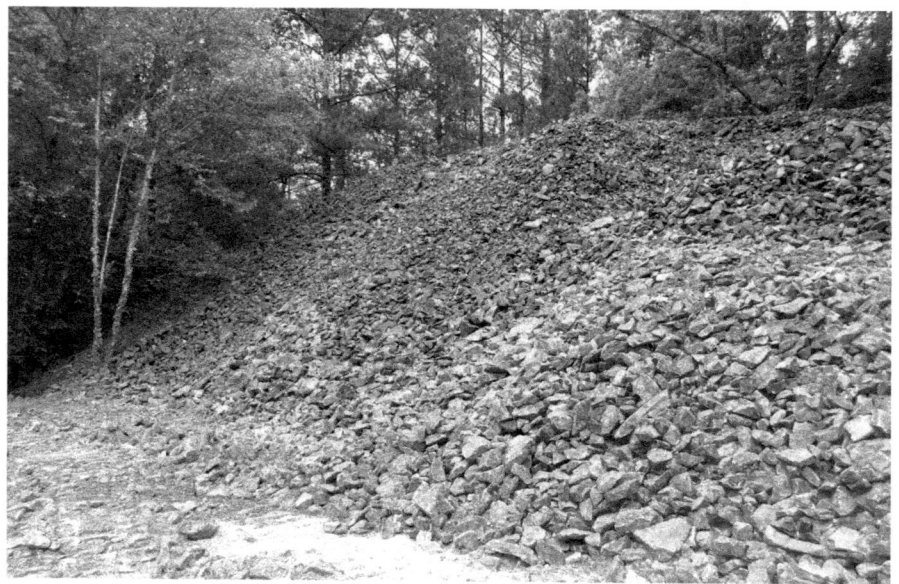

The Hog Mountain rock quarry was composed of ore of low grade or no value that was discarded from the mill. *Courtesy of Peggy Jackson Walls.*

Before moving to the Hog Mountain area, Mr. Neil owned a jewelry store in Ohio. During the Depression, he lost all of his money in the stock market crash and was completely broke when he became caretaker at Hog Mountain, where he and his wife lived for ten to twelve years after the mine closed. He was in his sixties at the time.

Mr. Neil and his wife arrived in their REO automobile, which soon fell into disrepair. Not having the money to fix the auto, Mr. Neil parked it by the house that he built on the edge of a drop-off near Hillabee Creek. He built a sluiceway from the sand pile to the house and would go up and put water in the sluice to wash the sand down to the house. He would sift through it looking for gold. Gold was cheap then, about thirty-five dollars an ounce. Mrs. Neil was a very petite lady, not weighing over 110 pounds. She loved cats and, at one time, had twenty-three at their home on Hillabee Creek. When the rolling stores went by, she bought PET milk by the case to feed her cats. She wore a scarf tied around her head all the time. After his REO was disabled, Mr. Neil had no means of transportation. The only time he left was when a rolling store came by or a storm damaged timber on Enitachopco. Then he would go over and clean up debris. Neighbors helped, taking timber as payment.

Surviving the Depression

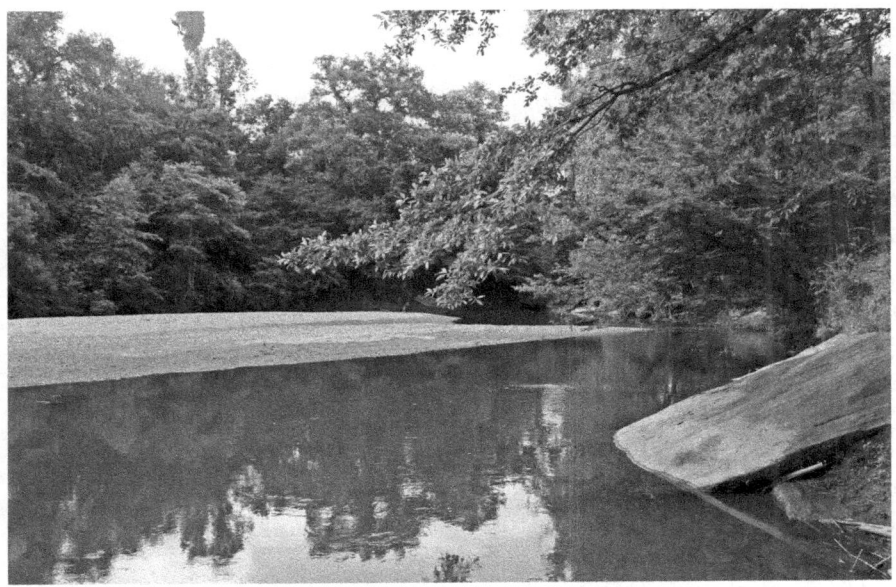

Hillabee Creek was a source of water for the Hog Mountain gold mine and a place for locals to try their hands at panning for gold. *Courtesy of Peggy Jackson Walls.*

Information about Mr. Neil was provided by Bernie Atchinson Jr. His father, Bernie Atchinson Sr., was sponsored by the Cravener family after leaving an orphanage. After completing two years at a junior college in South Carolina, Bernie came to the Hog Mountain area and stayed with the caretaker, Mr. Neil, and his wife for about a year after the mining operation had closed.

Bernie lived with the Neils for about a year before he went into military service. Mr. Neil wrote to Bernie telling him about his wife's death. By the time the Craveners received the message and could return to Hog Mountain, she was already buried. After the service, Bernie went to Auburn. He returned to Hog Mountain one day to visit Mr. Neil and found he had closed the house and left. His departure was as mysterious to Bernie as it was to members of the Hog Mountain community.

Alabama Gold

Otis Young

Otis Young lived about a mile from Hog Mountain and worked on his farm until he learned the mine was reopening. Like other men in the Hog Mountain communities, he welcomed an opportunity to earn a regular payday. When he applied for a job, he quickly got one.

When I worked there, George Brown was the superintendent; Colley was the foreman; Davis was the boss over the second shift...We did the drilling two hundred feet deep and one-half mile north and one-half mile south. There was one accident, happened to Bonnie Reed. He was pulling muck out of place, and it caved in. He went down in there about one hundred feet. We worked about four hours getting him out, and he lived. He continued to work at the mine, but he didn't do work as a mucker any more. Jimmy Farrow run the hoisting machine. I helped Walter Brown haul ore from the shaft to the conveyor [and] from there to the ball mill, where they ground it up.

I walked about a mile to the mine until I got a car, a '31 A-Model Coupe...They had a commissary there with everything, cigarettes and eats. We had a mess hall and cook, and a lot of folks stayed there and ate there all the time. Mostly, I carried my lunch. There were rolling stores that come by the mine—Hester Eason and Cecil Edwards at Eagle Creek. Wiley Bennett run a store in Cowpens.

I was working nine hours a day in the mine when they went to taking out Social Security in '36. They said the mine was going to close. There weren't many jobs then....My daddy-in-law run a sawmill. They sawed logs with crosscut saws back then, one man on each end. Didn't pay as much as the mine. So I got a job building metal caskets for two years. After that I went to Russell Manufacturing Company and put in for a job, and they called me. People who come there to start the mine were from Tennessee and Birmingham. They were already experienced hands. Then they'd train the workers.

Alton Padgett

Alton never worked in the mine but recalled his elder brother Elbert's stories about working at Hog Mountain mine. Elbert lived with an aunt and walked across the mountain from her house to the mine six days a week.

Surviving the Depression

He and Jesse Lovelady worked together separating rock. Miners would dig until they could see tree roots and then stop to keep the tree from falling in on them. When Elbert left the mine, he worked for Russell Mills in the weave shed until he retired.

Lynwood Champion

My name is Otis Lynwood Champion. My parents were Otis Champion and Ceiley Osborn Champion. We lived in Tallapoosa County all our lives except for two years when Dad worked in the oil field in Texas. We came back to Tallapoosa County and bought a farm, where I was farming when the mining started up. I was about eighteen when I went to work in the blacksmith shop with my wife Luveria's father, sharpening drill steel.

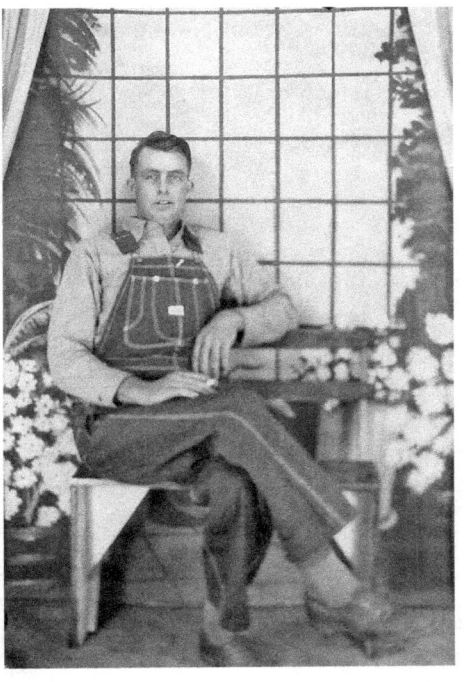

Hog Mountain gold miner Jesse Lovelady worked with Elbert Padgett separating rocks and ore in the 1930s. They earned a one dollar a day for their labor. *Courtesy of Lynn Lovelady Willis.*

When we first got married, we lived with Luveria's brother Carey White. Then we moved to this farm about two and one-half miles from the Hog Mountain. George Brown was first superintendent of the operation, and Mr. Collins was the last assistant superintendent—mighty good men. Price McLeod and Bill Whiting worked in the shop with Roy Yates, Otis Peppers and Carey White.

Hayes Simpson helped to clear a throughway for the electric lines when they moved the heavy equipment up the mountain. Hayes, Buddy McGhennis and Theodore White were assistant electricians. J.P. Mooney and Howard Mooney both drilled in the mine.

Some of the workers stayed in the old hotel that was still standing from the operation that closed in 1916. It was built before World War I, and it was a two-story structure built of heart pine. They just put a new roof on the hotel

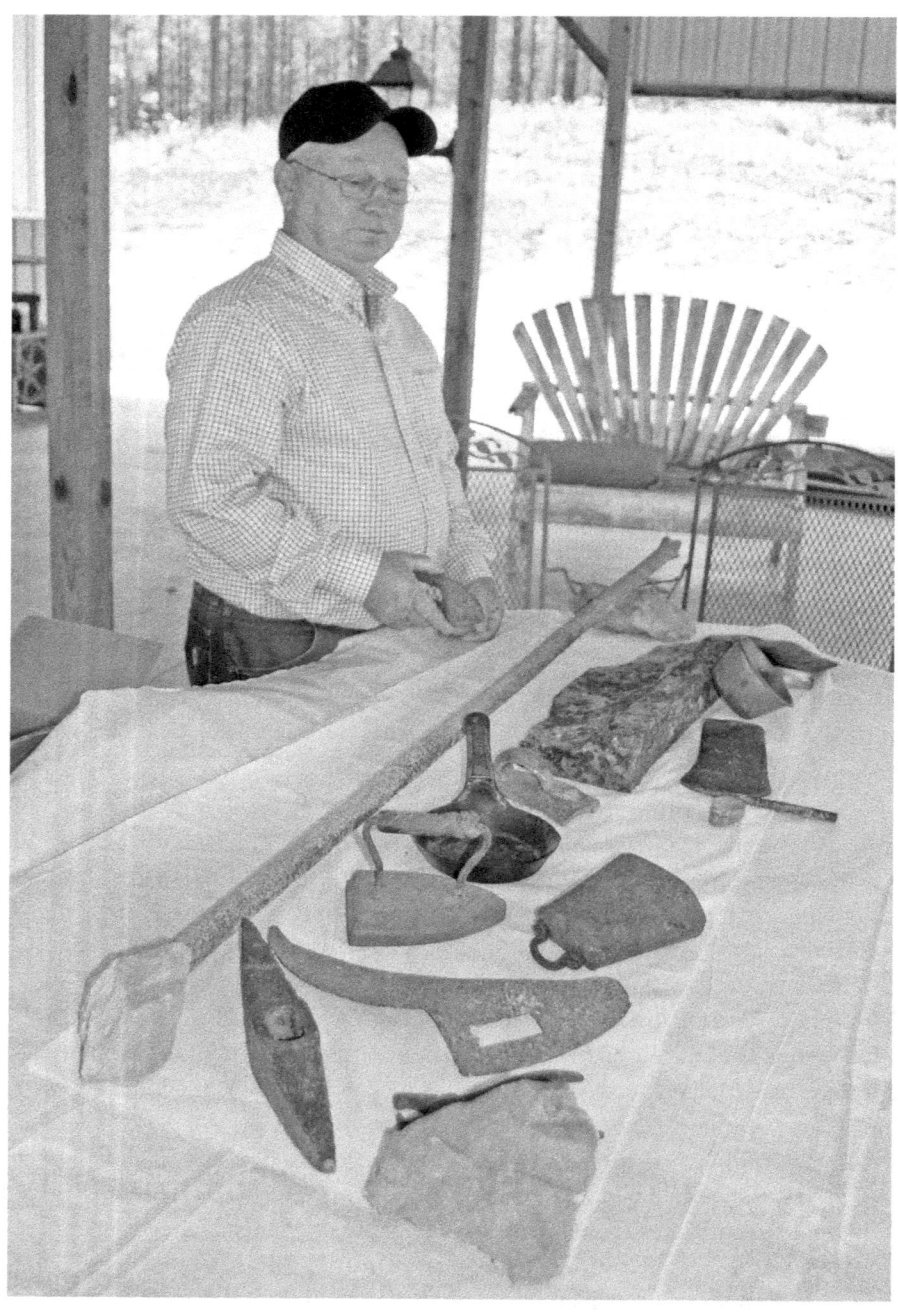

Douglas Champion was the son of machinist Lynwood Champion and Luveria, whose father, Carey White, operated the machine shop in the 1930s operation. *Courtesy of Peggy Jackson Walls.*

and used it for a bunkhouse. Just a place for them to sleep. But they ate at the mine. There was another bunkhouse with five or six rooms. There were four old houses on Chicken Row in a straight line, southwest from the mine. Buddy McGhennis lived in one of them and Hershel Peppers in another. They told us that we would be out of a job in a certain length of time. I had a mule, so I quit and started plowing the fifteenth of April. Later I got a job at Avondale Mills, where I worked for forty years and never missed a payday even when I lost my hand working on a big ole gear that had spokes in it that turned slow speed. It turned close to a big ole air duct with a little door on it that would come off.

Luveria Champion

I'd like to have a history of that place. I had two brothers, a daddy and a husband to work over there. We had an old '28 or '29 Chevrolet, but most of the miners walked. There were trails all over the mountain where people walked. Some of them talked about how scary it was coming through there at night. Plenty of holes to step into if you weren't careful. There were bobcats that sounded like a woman screaming.

When the bridge fell in and that Negro man was killed, we were living with Carey and his wife, Freddy. The driver came running up to Carey's house and straight in the door, didn't even stop to knock. He was bloody. He said, "Somebody carry me to the mine, and somebody go stop the traffic. The bridge fell in, and it killed the black man." Well, I couldn't have drove the car. I told him, "I'll give you the keys and you can drive our car." And I ran all the way to stop the traffic. When I got there, the insurance man was standing in the bridge. He said, "I pumped the brake to stop." He got right up to where the hole was: "I pumped the brake, and I pumped them again and stopped right just as I got to the hole." Scared him so. So he stood on this side of the creek, and I was on the other side of the bridge when Mr. Hershel Peppers came flying. It wasn't fast like you'd drive now but was fast at that time. And I was out there just a-flagging. His wife's mother, Mrs. Self, got out, and she said, "Luveria, I'll always love you. If you hadn't been out there, we'd went right into that hole before we ever got stopped." You couldn't see from the outside of the covered bridge that the bottom was out. When they hauled that concentrate, it was heavy. That's what made the bottom fall out. The water wasn't deep, but I imagine something fell on him and killed him. It was a long time before they got the bridge repaired.

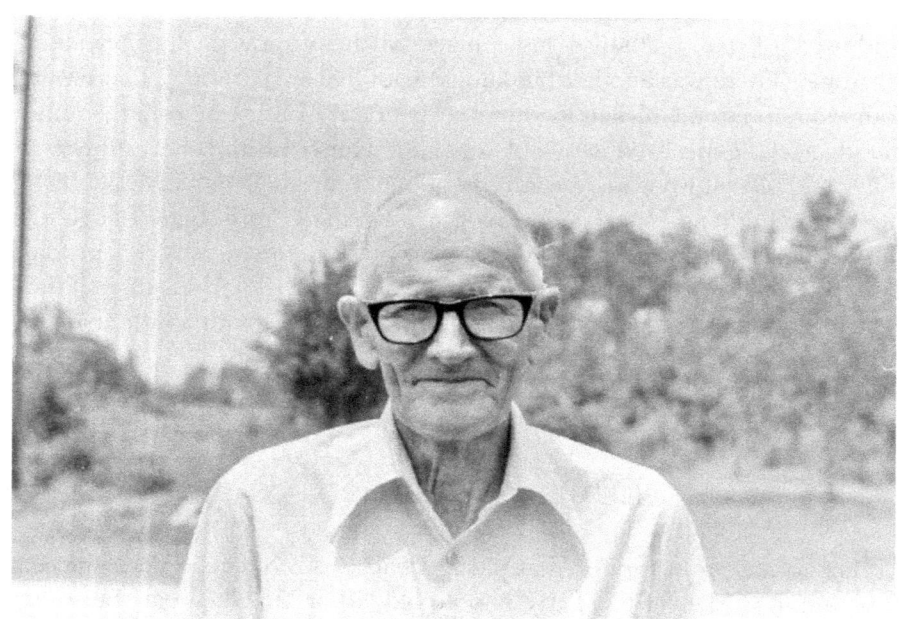

Benjamin Talmadge Jackson was a Hog Mountain gold miner in the 1930s. He worked as a mucker and operated tram cars. *Courtesy of Peggy Jackson Walls.*

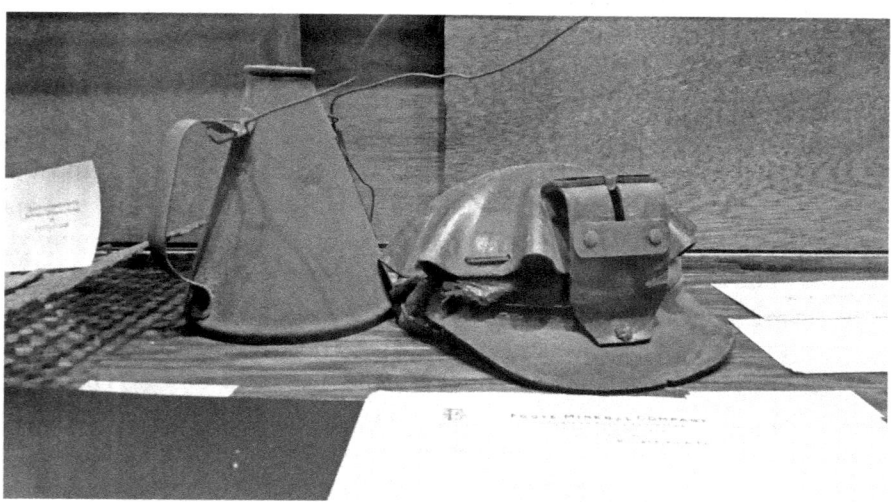

This picture of a miner's helmet is used with permission of Pine Mountain Gold Museum, located in Villa Rica, Georgia. *Courtesy of Peggy Jackson Walls.*

Surviving the Depression

Charlie Thomas Scott

I worked at Hog Mountain for a year, mostly timbering. When blasting was done, there was a lot of loose rocks that had to be timbered. My brothers Pres and Olan worked there for a while, too. Olan like to have blowed himself up when he was helping ole Charlie Worthy with the dynamite charges. You were supposed to fill them and then light them. Charlie was filling the charges, and Olan came alone behind him, lighting them with the light on his cap. Charlie turned around and saw what Olan was doing. He said, "What the hell are you doing? Get out!" and they run. You see, they would run wires out from the charges, and before those wires were touched, Charlie would get way down around the bend in them tunnels. But Olan had decided to help and come right along behind him lighting the fuses. They had to go quick! We worked five days a week and went back to the mine on Saturday to pick up our pay.

Talmadge Jackson

Charlie and I help[ed] build all the houses with Rob Bowens, the head timberman. His helpers were his son Aubrey, Joseph Cleveland, Ed Walls, Talmadge Jackson and me. Then I worked as a mucker inside the mine. I had a couple of accidents while I was working there. One was caused by the way those tram cars operated before they put mechanical brakes on them. Well, the brakes were nothing but planks pulled out there on the car, and you sat down on that plank and used your weight for the brake. When the tram car was loaded, I'd ride the car and the plank out. One time I was riding it, and it got to going so fast that it got away from me. I hit upside the rocks, and it knocked me off. Knocked me out for a little while. After my accident, Ed Walls installed mechanical brakes on the tram.

The other accident I had was a little more serious. A carbide can blowed up in my face and took all the skin from around my eyes. It had been raining that night, and this carbide had been sitting out and gotten damp. I walked too close to the can [and] the light on my cap set it off. There wasn't a car at the mine that night, and Mr. Brown was gone in his. So we started out on a skidder to go to the doctor. It was pouring rain. We met Mr. Brown coming in and got out of his car and let us have it. They carried me on to the doctor. The next day, a lawyer tried to get me to sue the company for $40,000. He said it wouldn't cost me a dime, that he would take the case for half of what

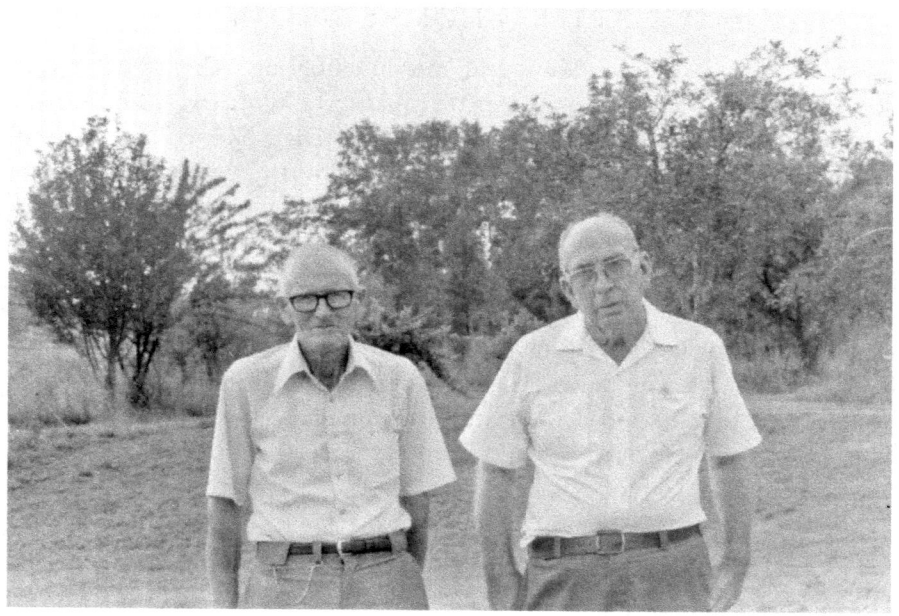

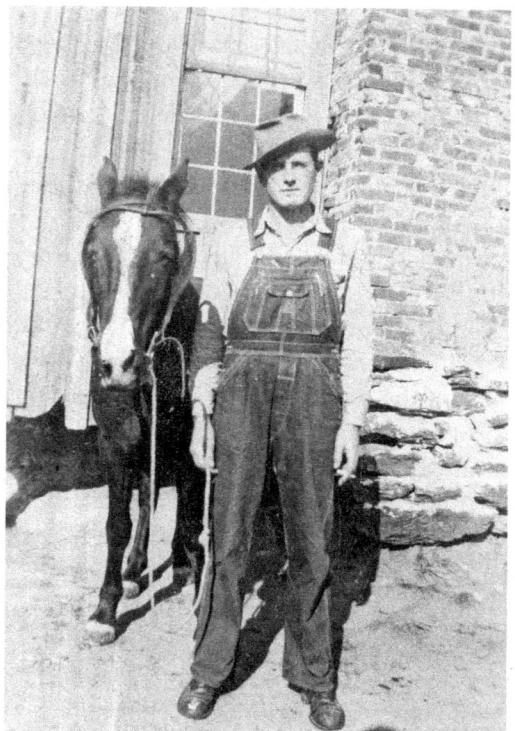

Above: Miners and friends Talmadge Jackson and Charlie Scott worked together at Hog Mountain during the Depression. *Courtesy of Peggy Jackson Walls.*

Left: Kermit Jackson with Kate, one of the two mules that worked at the mine in the 1930s. *Courtesy of Peggy Jackson Walls.*

Surviving the Depression

we won. I told him I wouldn't do it. Mr. Brown had been too good to me. He paid my doctor bill and expenses, paid me my regular time right on while I had to be out.

When I was working as a mucker, a little German helped me. Some of the workers came from far off. They put air pipes in the mine while I was there. They didn't have them to start with when my brother Kermit and some of the first ones went to work there. And when they blasted, they didn't have no way of blowing smoke out of there. When they did put the air pipes in; it helped some, but they still had dust in there. That was why so many of the boys developed silicosis: Kermit, Marvin Dean, Johnny Coker, Cecil Osborn and the Watley twins. They all died young. I don't know for sure when I got hard of hearing, whether it was while I was working in the mine or not. It just gradually come on me. I laid it to that racket in the weave shed when I worked at the mill. I had to wear ear plugs all the time in the mill.

Talmadge grew a variety of vegetables on his farm that he took to market in Alexander City. In season, he had watermelons, cantaloupes, peppers, tomatoes, cabbage, potatoes and green peanuts. He and other truck farmers had an assigned area across from Avondale Mills where they could park their trucks in the shade to keep their products fresh. After millworkers finished their shift, they could pick up fresh vegetables as they started home, and the truck farmers could add a few dollars to their retirement fund.

Emra Allen

Work had just started when I got a job digging drainage ditches with Thomas Daugherty [in] 1931. My next job was firing the boiler that operated air drills inside the mine 'til a gasoline engine was installed. It was used until lines were run from the Alabama Power plant in Alex City. Then, electricity was used to run the plant. My next job was firing the boiler. There were just six of us working the night shift. Thomas Daugherty run the hoisting machine; the driller, his helper and two muckers were working below. I didn't work but seven shifts on top before I went to work underground as a mucker with George Dean. They paid a dollar a day for muckers then.

I worked as a mucker until they needed more drillers. Then I learned how to operate the drills, the Sullivan and the Jackhammer. Mr. Green would work half the night to get water running into that drill. The old-timers used to say, "Always get your water to running. Never drill with a dry drill." A lot

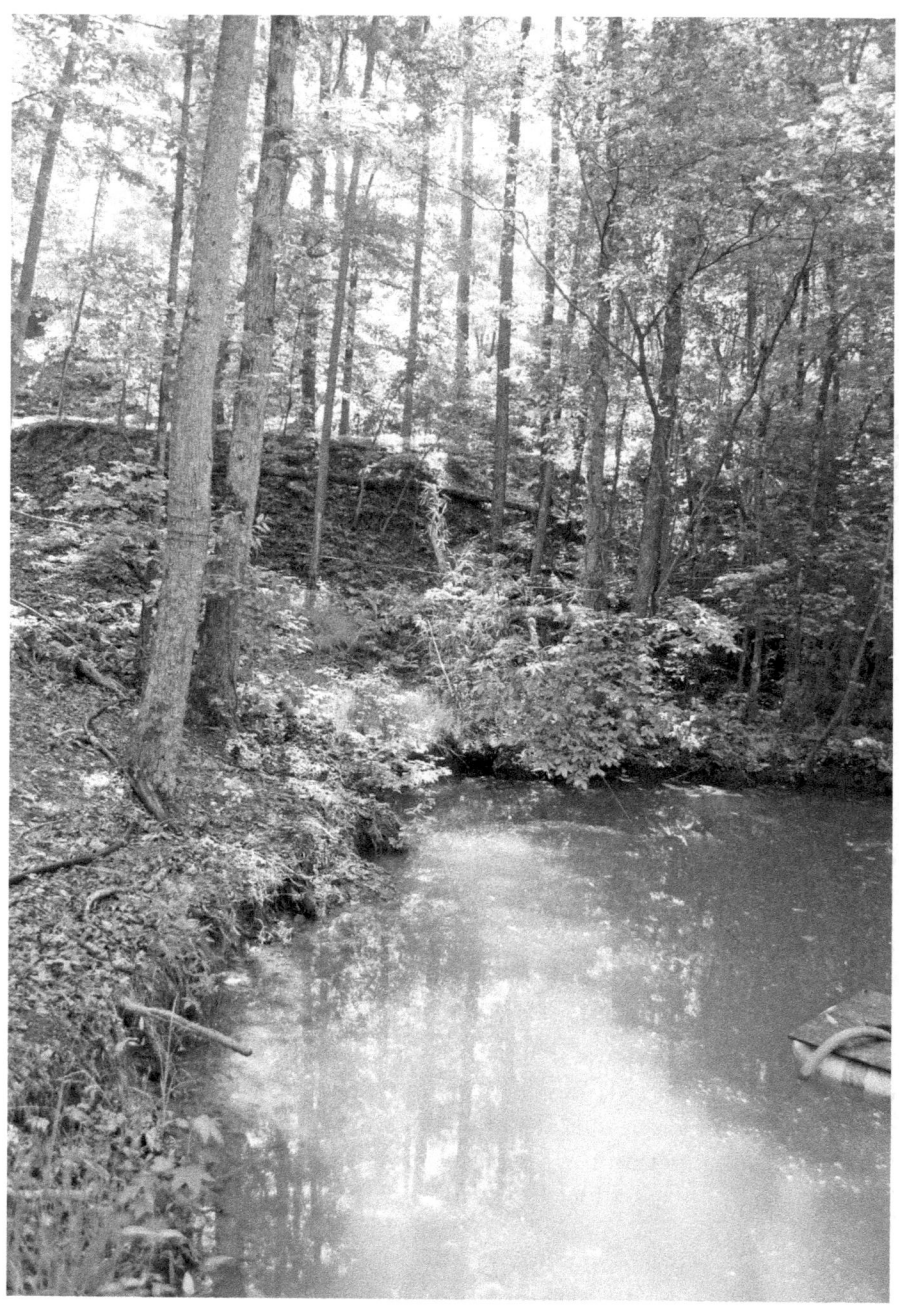

The Hog Mountain tailing pool was near the location of the old mill structure and the rock quarry. *Courtesy of Peggy Jackson Walls.*

Surviving the Depression

of the young fellars never worked in the mine before, and they didn't pay no attention to what they said and used dry drills because they could drill a little faster. A lot of them are dead now from silicosis—Rhett McWright, Cecil Osborn…When Cecil would come out of the mine, he'd look like he come out of a powder box with dust all over him.

Sometimes, we would drill close to the old headings made in the last mining operation, and there would be a little drill hole. The old-timers Mr. Green and Mr. DeFord told me, "Never drill under that hole. Always drill to the side of it or above it. If they left a little powder that nitric glycerin would work right down, and you're liable to drill in there and get killed.

The mine captain on my shift was Mr. Dye. I worked for Mr. Green and Mr. DeFord, too. Other workers were Travice Foster; he laid the track in the mine. My brother Warner worked with me as a driller. My other brother John worked in the mine one shift. He was down in the mine that morning when they got to shooting, and the whole mine shook. He said, "You're crazy, boy. I wouldn't work in that place." He just worked that one shift.

When I worked at the mine, I boarded some at Mr. Jim Jones. When I stayed at the camp house, they took out of my pay for board and meals. The old cook and his wife, a young woman, cooked three meals a day on a wood stove. Now some of the men brought lunches with them and just ate in the mine 'cause there was always water to wash your hands with down there.

There wasn't much time for recreation, but occasionally, we'd play cards in the camp house. Sometimes, we'd go to Alex City to see one of them westerns and that singing cowboy Roy Rogers. I left the mine in 1936 to join the army and went to the Panama Canal Zone.

Emra served in the army for four and a half years before returning home. He worked at "the cotton mill" in Alex City and then retired to his farm at Cowpens. Emra's story is similar to other Hog Mountain gold miners who left the mine for a job in the mill and served their country during World War II. They had confidence that "Hog Mountain would run again" when gold prices were high and labor was cheap. It is likely they did not realize they were a part of a historic event, when in 1936, the mine yielded 4,726 ounces of gold, valued at $165,410, making Alabama the leading producer of gold in the Appalachian states.[75]

From the Mine to the Mill

J.P. Mooney

When I went to work at Hog Mountain, I stayed with the Claude Woodruff family at Cowpens. It took an hour and half each way to walk from Woodruff's place through the woods to the mine. My brother Howard and I worked cleaning out the old shaft before Simon and Cowen Construction Company lowered the shaft to below the two-hundred-foot level. That was in 1933. Before that, I worked on the top house, helping to unload rock and debris that came out of the mine. After that I worked as a mucker, making $1.25 for a nine-hour shift. That went up to $2.75 before the mine shut down. Believe me, that's impressive work, a little backbreaking. Then I

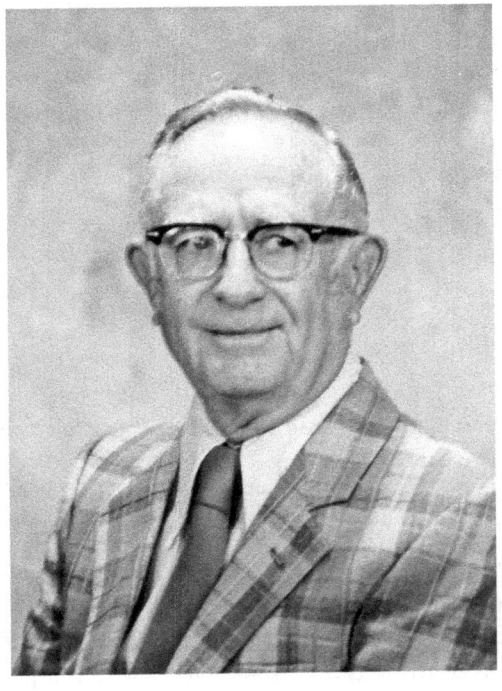

Hog Mountain miner J.P. Mooney later served on the national committee that set standards for mine safety. *Courtesy of Peggy Jackson Walls.*

graduated to more skilled jobs. First, I became a drill helper [and] then a driller. For the last two years I worked there, I was head of the blasting crew on the third shift.

While I was working there, I helped lead the work stop that resulted in the company's giving us a five-cent-an-hour increase in pay, which amounted to a forty-five-cent raise for a nine-hour shift. Some of us approached management on a Friday with the request that the third shift would be rotated. We were told that they would make that decision and that we had nothing to do with it. So I got a lot of signatures on a petition and served as spokesman. When we presented it, we were told, "You'll come to work when you're told." The entire nightshift didn't go to work Saturday or Sunday. When we got there Monday, we were told the day shift was going to work. We took a stand and told them "Not if we have anything to do with it." We went to the shaft and got some pieces of steel and told them, "Anybody works, we're going to work. We're going to have some rotations of shifts, or we're going to have a fight."

We were trying to deal with the underground superintendent, Mr. Beavers, a stubborn old gentleman. When they got Mr. Brown, the general manager, out, he agreed with us there should be some rotation so that no one would have to do all of the nightshifts. About two weekends later, they did change the shifts. We didn't have a union, but we organized to get rotation of shifts and a wage increase. Mr. Brown had worked as a laborer at some time and was sympathetic with our requests. Some of the men I worked with in the mine were Rhett McWright, Luther Dye, Mon Woodruff, Albert Bence, Orville Bence, Fay Melton and my brother Howard. Most of them died with silicosis caused from inadequate ventilation. Precautions that were taken were to put some vents in to pipe air into the working areas. But this was done only on one level. There was no way to ventilate the second level. So many of those raises only had one way up, and workers had to come down the same way. It would have been expensive for the company to have a second entrance-exit. What resulted was if anybody got closed off, he was trapped. And one man did. The timbermen were putting up a scaffold for the driller to work off. A man fell and was buried up to his neck. He tore all of his fingernails off, trying to scratch his way out. They set some timbers in and put up chain blocks and tackles around a miner's waist so he could go down and dig out around the trapped miner. Eventually, they were able to put some ropes under his arms and pull him out of the muck pile. That ended his mining career right there. And I became interested in mining safety and working conditions.

Surviving the Depression

After the mine closed, I worked at Ramer mine near Bessemer. There was an effort being made to organize the workers in a union. In 1939, I joined the International Union of Mine, Mill and Smelter workers at the steel mills around Ensley and Fairfield. The last year I was there, I served as staff representative of the union. Then when Ramond mine closed in 1943, I went to work for the United Textile Workers of America.

MOONEY BECAME INVOLVED IN what proved to be an unsuccessful effort to unionize textile mills in southeast Alabama, in Alexander City, Sylacauga and Opelika. He eventually had the opportunity to work with the union for better safety and health conditions for workers in mines, smelters, refineries and mills. He was recognized for his work with two appointments to an advisory committee, established in 1967, to recommend proper standards for the enforcement of the 1966 Federal Metal and Nonmetal Mine Health Safety Act. As safety coordinator for fourteen western states, J.P. worked for the passage of the Federal Mine Safety and Health Act of 1977. After retiring, J.P. served on a nine-man Blue Ribbon committee to compile information on the research and development of communication systems by the US Bureau of Mines.

The closing of the Hog Mountain Mining and Milling operation in 1937 marked the end of major mining activity in Alabama in the twentieth century. But as recreational miners and descendants of the old-timers continue to dip their gold mining pans and look for "color" in the creeks of Hillabee, Enitachopco and Broken Arrow, the story and tradition of gold mining in Alabama will continue.

Appendix

An Incomplete List of Hog Mountain Gold Miners

Emra Allen	mucker	Tucker Cash	mucker
Warren Allen	mucker	Lynwood Champion	machine shop
Lander Baker	team of horses	Byron Cleveland	carpenter
Walter Neal Barfield	driver	Heflin Cleveland	carpenter
Mr. Beaver	manager	Joseph Cleveland	carpenter
Albert Bence	mucker	John Coker	timberman
George Biggs	superintendent	Mr. Colley	foreman
Bennie H. Bonner	carpenter	Thomas Daugherty	hoist man
Hershel Bonner	mucker	Mr. Davis	foreman
John Bonner	carpenter	Cevial Dean	picker
Marshal Bonner	mucker	Clark Dean	driller
Aubrey Bowen	carpenter	Morgan Dean	mucker
Marion Bowen	stoper	X Dean	driller
Rob Bowen	carpenter	Mr. DeFord	shift captain
Jim Britton	boiler	Frank Downs	sand bed
Doc Brown	driller	Abner Duke	mucker
Frank Brown	picker	Ace Duke	picker
George Brown	manager	Mack Duke	driller
Thomas Brown	hoist man	Ronnie Duke	timber
Walter Brown	carpenter	Mr. Luther Dye	shift captain
Hardy Buckner	driller	Aber Earl	stopper
Clarence Butler	watchman	Jimmy Farrow	mucker

Appendix

Travice Foster	cleared land	Howard Mooney	driller
Woodson Galloway	mucker	J.P. Mooney	driller and explosives
Alvin Goodwin	driver		
Carl Green	assay work	Jim Nelson	driver
Harry Wallace Green	assay work	Mr. Ogles	canned concentrate
Llewellyn Green	assay office	Mr. Oplin	supervisor
Willie Green	assay office	Cecil Osborn	driller
Charlie Harris	driller	Cowboy Osborn	mucker
John Harris	sand bed	Elbert Padgett	picker
Bond Jackson	mucker	Buck Patterson	timberman
Kermit Jackson	driller and timberman	Hershel Peppers	carpenter
		Bonnie Reed	mucker
Talmadge Jackson	operated tram cars	Charlie Scott	timberman
		Hayes Simpson	electrician
Will Jarvis	hauled machinery and cleared land	Frank Smith	mucker
		James W. Steward	mucker
Eugene Leslie	timberman	Marshal Edwin Walls	chemist
Jesse Lovelady	picker	Bennie White	machine shop
Grant Lowe	mucker	Carey White	machine shop
Ed Mahan	picker	John White	blacksmith shop
Fey Melton	driller	Richard White	machine shop
Herschel McCllelan	mucker	Theodore White	electrician
Buddy McGhennis	electrician	Bill Whiting	machine shop
Price McLeod	machine shop	Frank Woodruff	driller
George McWhorter	mucker	Mon Woodruff	driller
James Henry McWhorter	driller	Charles Worthy	mucker
		William Worthy	mucker
Rhett McWright	drillers	Otis Yates	mucker and driller
Lucius Monroe	driller	Roy Yates	mucker

Rolling stores: Red Eason, Lafayette Evers, Hester Allen
Boardinghouses: Miss Daisy Simpson, Miss Eula Green, Rob Bowen, Jim Jones and Claude Woodruff

Notes

1. Weddle, "European Exploration."
2. Jefferson, "Query VI."
3. Sweet, *Gold in Virginia*, 1–2.
4. Myers, "North Carolina."
5. "History of Gold."
6. "Exhibit: Haile Gold Mine."
7. "Georgia Gold Rush."
8. Williams, *Georgia Gold Rush*, 22–23.
9. Phillips, "Preliminary Report," 97.
10. Adams, "Gold Deposits of Alabama," 7.
11. Phillips, "Preliminary Report"; Adams, "Gold Deposits of Alabama"; Williams, *Georgia Gold Rush*.
12. Braund, "Continuation of the War of 1812."
13. Jensen, "Battle of Horseshoe Bend."
14. "People & Events: Indian Removal 1815–1858."
15. *Niles Weekly Register*, "Description of the gold mines," 369 in Dean, "Golden Harvest of the Piedmont," 23–24.
16. Dean, "Golden Harvest of the Piedmont," 23–24.
17. Ibid., 23.
18. Waters, "Goldville District."
19. Tuomey, "Second Biennial Report," 292.
20. Johnson, "Mining and Milling Methods."
21. Phillips, "Preliminary Report."

Notes

22. Russell, "History of Benjamin Russell."
23. Aldrich, Hillabee Gold Mine Prospectus.
24. Russell, "Gold Mining in Alabama," 5–14.
25. Phillips, "Preliminary Report," 41.
26. Phillips, James D., to Sarah Ann, Tallapoosa Historical Museum, Dadeville, AL.
27. Kidd, "History of Blue Hill."
28. Young, "Southern Gold Rush," 391–92.
29. Brannon, "Robert Grierson," 818.
30. Ibid., 819.
31. Robert Grierson to Benjamin Hawkins, September 23, 1813, "Letters of Benjamin Hawkins, 1797–1815," edited by J.E. Hays, Georgia Department of Archives and History, 254.
32. Brannon, "Robert Grierson," 818–20.
33. Saunt, Claudio. *Black, White, and Indian*, 39.
34. Kappler, "Treaty with the Creeks."
35. Saunt, from Grayson, *Creek Warrior for the Confederacy*.
36. Saunt, "Creek Indians."
37. Saunt, from Grayson, *Creek Warrior for the Confederacy*.
38. Hooper, *Adventures of Captain Simon Suggs*.
39. Ibid.
40. Hoole, *Alias Simon Suggs*.
41. Soloman, "Simon Suggs."
42. Dean, *Papers of Michael Tuomey*.
43. Ibid.
44. Ibid.
45. "Cotton Economy in the South."
46. Copeland, "Hooper."
47. Reynolds, "Reluctant Rebels," 88.
48. Spence, "Real Rhett Butler Revealed."
49. Kuenzi, "Search for the Lost Confederate Gold."
50. "Private Mint and Territorial Gold."
51. Adams, "Gold Deposits of Alabama," 8.
52. "Rogan plate."
53. *Niles Weekly Register*, July 17, 1830, 320–33.
54. Adams, "Gold Deposits of Alabama," 9, 10.
55. *Vidette*, circa 1888.
56. Ibid.
57. *Alexander City Centennial, 1874–1974*.

NOTES

58. Ibid.
59. Ibid.
60. Ibid.
61. Phillips, "Preliminary Report," 41.
62. Ibid.
63. Farrow family records, courtesy of John F. Fletcher.
64. Coley, "Climax of Gold Mining in Alabama."
65. Farrow family records, courtesy of Tallapoosee Historical Museum, 4.
66. Brewer, "Preliminary Report."
67. Dean, "Minerals of Alabama."
68. Adams, "Century of Gold Mining in Alabama," 271–79; ibid., "Gold Deposits of Alabama," 91.
69. Phillips, *Geological Survey*, 40–50.
70. Aldrich, "Treatment of the Gold-Ores," 578–83.
71. Ibid., Hillabee Gold Mine Prospectus.
72. Downs, "Great Depression in Alabama."
73. Adams, "Century of Gold Mining," 278.
74. Coley, "Climax of Gold Mining in Alabama."
75. Simpson and Neathery, "Alabama Gold," 57.

Bibliography

Adams, George I. "A Century of Gold Mining in Alabama" *Alabama Historical Quarterly* 1 (Fall 1930): 271–79.
———. "Gold Deposits of Alabama and Occurrences of Copper, Pyrite, Arsenic, and Tin." *Alabama Geological Survey Bulletin* 40, 1930: 8.
Aldrich, T.H., Jr. Hillabee Gold Mine Prospectus, circa 1904. Alabama Department of Archives and History. Montgomery.
———. "The Treatment of the Gold-Ores of Hog Mountain, Alabama." *American Institute of Mining Engineers Transactions* 39 (1909): 578–83.
Brannon, Peter. "Robert Grierson, Trader at Hillibee Town." *DAR Magazine*, October 1949.
Braund, Kathyrn. "Continuation of the War of 1812." http://www.encyclopediaofalabama.org/article/h-1820.
Brewer, W.M. "A Preliminary Report on the Upper Gold Belt of Alabama in the Counties of Cleburne, Randolph, Clay, Talladega, Elmore, Coosa, and Tallapoosa." *Alabama Geological Survey Bulletin* 5 (1896): 1:1–105.
Coley, Judge C.J. "The Climax of Gold Mining in Alabama." Lecture. AHA State Conference. Mobile, AL, May 5, 1967.
Copeland, J. Isaac. "Hooper, Johnson Jones." 1998. NCpedia.org. http://ncpedia.org/biography/hooper-johnson-jones
"The Cotton Economy in the South. *American Eras*. 1997. Encyclopedia.com. http://www.encyclopedia.com/doc/1G2-2536601340.html
Dean, Lewis. "Minerals of Alabama." 2007. *Encyclopedia of Alabama*. http://www.encyclopediaofalabama.org/article/h-1273.

Bibliography

Dean, L.S. "Golden Harvest of the Piedmont." *Alabama Heritage*, Summer 1991: 21–28.

———. "Michael Tuomey's Reports and Letters on the Geology of Alabama, 1847–1855." *Alabama Geological Survey* open-file report, 1986: 80.

———. *The Papers of Michael Tuomey*. Spartanburg, SC: Reprint Company, 2001.

Downs, M.L. "Great Depression in Alabama." 2014. *Encyclopedia of Alabama*. http://www.encyclopediaofalabama.org/article/h-3608.

Emerson, E.H. "The Hillabee Mine, Hog Mountain, Alabama." *Alabama Geological Survey* unpublished report, Pamphlet File 11, 1934: 7.

"Etowah plates." Wikipedia. https://en.wikipedia.org/wiki/Etowah_plates.

"Exhibits: Haile Gold Mine." Lancaster Public Library: Local History & Genealogy. http://lanclib.org/history/resources/exhibits/hgm/hgm.htm

Garner, George. "Gold Production in Alabama." *Encyclopedia of Alabama*. http://www.encyclopediaofalabama.org/article/h-1666.

"Georgia Gold Rush." AboutNorthGeorgia.com. http://www.aboutnorthgeorgia.com/ang/Georgia_Gold_Rush.

Grayson, G.W. *A Creek Warrior for the Confederacy: The Autobiography of Chief G.W. Grayson*. Norman: University of Oklahoma Press, 1991.

Green, Fletcher, M. "Georgia's Forgotten Industry: Gold Mining." *Georgia Historical Quarterly* 19 (1935): 91–111, 210–28.

"The History of Gold in North Carolina." *Gold Fever and the Bechtler Mine*. UNCTV.org. http://goldfever.unctv.org/history.

Hoole, W. Stanley. *Alias Simon Suggs: The Life and Times of Johnson Jones Hooper*. Tuscaloosa: University of Alabama, 1952, x.

Hooper, Johnson Jones. *Adventures of Captain Simon Suggs*. Tuscaloosa: University of Alabama Press, 1993.

———. *Some Adventures of Captain Simon Suggs, Late of the Tallapoosa Volunteers*. Philadelphia: Carey & Hart, 1845.

Jefferson, Thomas. "Query VI." *Notes on the State of Virginia*. Philadelphia: Prichard and Hall, 1781. *Documenting the American South*. UNC.edu. http://docsouth.unc.edu/southlit/jefferson/jefferson.html#p24.

Jensen, Ove. "Battle of Horseshoe Bend." *Encyclopedia of Alabama*. http://www.encyclopediaofalabama.org/article/h-1044.

Johnson, N.O. "Mining and Milling Methods and Costs, Hog Mountain Gold Mining and Milling Company, Alexander City, Alabama." *US Bureau of Mines Information Circular 6914*, 1935, 23.

———. "Mining methods and ore estimation at the Hog Mountain mine." *American Institute of Mining, Metallurgical and Petroleum Engineers Transactions* 126 (1937): 34–45.

Bibliography

Kappler, Charles J., ed. "Treaty with the Creeks, 1832." *Indian Affairs: Laws and Treaties.* Vol. 2, *Treaties.* Washington, D.C.: Government Printing Office, 1904. http://digital.library.okstate.edu/kappler/vol2/treaties/cre0341.htm.

Kidd, Ben E., III. "History of Blue Hill." Kidd family records. Tallapoosee Historical Museum, Dadeville, AL.

Kuenzi, Hans. "The Search for the Lost Confederate Gold." *Charger* 29, no. 7 http://clevelandcivilwarroundtable.com/articles/chargers/08/charger0308.pdf.

Myers, Caron. "North Carolina: The Golden State." *Our State.* http://www.ourstate.com/north-carolina-gold-rush.

Neilson, Mike. "Piedmont Upland Physiographic Section." 2007. *Encyclopedia of Alabama.* http://www.encyclopediaofalabama.org/article/h-1309#sthash.QM2R0bOf.dpuf.

Niles Weekly Register. "Description of the gold mines in Georgia and the Cherokee nation." July 17, 1830, 369. http://babel.hathitrust.org/cgi/pt?id=pst.000055571425.

Park, C.F., Jr. "Hog Mountain Gold District, Alabama." *American Institute of Mining and Metallurgical Engineers Transactions, Mining Geology* 115 (1935): 209–28.

"People & Events: Indian Removal, 1815–1858." Resource Bank. *Africans in America.* Part 4, *Judgment Day: 1831–1865.* PBS.org. http://www.pbs.org/wgbh/aia/part4/4p2959.html

Phillips, W.B. "A Preliminary Report on a Part of the Lower Gold Belt of Alabama in the Counties of Chilton, Coosa, and Tallapoosa." *Alabama Geological Survey Bulletin* 3 (1892): 97.

"Private Mint and Territorial Gold of the American 19th Century." Bunker Hill Rare Coin. http://www.bunkerhillrarecoin.com/?id=17.

Prospectus for Hillabee Gold Mine Company. Alabama Department of Archives and History.

Reynolds, Gerald H. "The Reluctant Rebels." Published by Tallapoosa County Bicentennial committee, 88–89.

A Ride with Old Kit Kuncker, and Other Sketches, and Scenes of Alabama. Tuscaloosa, AL: M.D.J. Slade, 1849. Reprinted as *The Widow Rugby's Husband.* Philadelphia, PA: A. Hart, 1851.

Robert Grierson to Benjamin Hawkins, September 23, 1813, "Letters of Benjamin Hawkins, 1797–1815," edited by J.E. Hays, Georgia Department of Archives and History, 254.

Russell, Ben. "The History of Benjamin Russell and Russell Lands Inc." 2009. BenRussell.com. http://benrussell.com/Ben-history%20of%20Russell%20Lands.htm.

Bibliography

Russell, R.A. "Gold Mining in Alabama Before 1860." *Alabama Review* 10, no. 1 (1957): 5–14.

Saunt, Claudio. *Black, White, and Indian: Race and the Unmaking of an American Family.* New York: Oxford University Press, 2005.

———. "Creek Indians." *New Georgia Encyclopedia.* http://www.georgiaencyclopedia.org/articles/history-archaeology/creek-indians.

Simpson, T.A., and T.L. Neathery. "Alabama Gold." *Alabama Geological Survey Circular* 104 (1980): 169.

Smith, E.A. "A General Account of the Character, Distribution, and Structure of the Crystalline Rocks of Alabama, and of the Mode of Occurrence of the Gold Ores." *Alabama Geological Survey Bulletin* 5 (1896): 108–11. http://www.gsa.state.al.us/documents/pubs/onlinepubs/Bulletins/Bull_5.pdf.

Soloman, Jack P. "Simon Suggs." In *Tallapoosa County: A History.* Edited by Bob Saxon. Alexander City, AL: Tallapoosa County Bicentennial Committee, 1976.

Southerland, Henry D., Jr., and Jerry Elizah Brown. *The Federal Road Through Georgia, the Creek Nation, and Alabama, 1806–1836.* Tuscaloosa: University of Alabama Press, 1989.

Spence, Dr. E. Lee. "The Real Rhett Butler Revealed." *Shipwrecks Blog.* http://shipwrecks.com/the-real-rhett-butler-revealed.

Sweet, Palmer C. *Gold in Virginia.* Mineral Resource Publication 19. Charlottesville: Commonwealth of Virginia Department of Conservation and Economic Development, Division of Mineral Resources, 1980, 1–2. https://www.dmme.virginia.gov/commercedocs/PUB_19.pdf.

Tuomey, Michael. *Information Series 77: Michael Tuomey's Reports and Letters on the Geology of Alabama, 1847–1856.* Edited by Lewis S. Dean. Tuscaloosa, AL, 1995.

———. "Second Biennial Report on the Geology of Alabama." *Alabama Geological Survey Biennial Report 2, 1858.*

Waters, Joe. "Goldville District." Gold Mining History Tallapoosa County. http://jovikri.tripod.com/public-index.html#Goldville District.

Weddle, Robert S. "European Exploration and Colonial Period." 2014. *Encyclopedia of Alabama.* http://www.encyclopediaofalabama.org/article/h-1180.

Williams, David. *The Georgia Gold Rush: Twenty-Niners, Cherokees, and Gold Fever.* Columbia: University of South Carolina, 1993.

Young, O.E., Jr. "The Southern Gold Rush, 1828–1836." *Journal of Southern History* 48, no. 3 (1982): 373–92.

Index

A

Adams, George I. 20, 21, 25
Age of Gold, the 45
Alabama Fever 29, 49
Alabama gold belt 9, 10, 25, 58
Aldrich, T.H., Jr. 21, 67, 85, 88, 94, 97, 106, 143
Aldrich, T.H., Sr. 67, 82, 97, 143
Allen, Emra 153
alluvial 30, 38, 46
antebellum 10, 11, 60
Appalachian Mountains 9, 15, 19, 20, 25, 60
Arbacoochee 32, 58

B

Bechtler family 18, 60
Birdsong Pits 34, 38
Black Belt 57
Blue Hill 20, 41, 45, 46, 76, 78, 79
Bowen, Rob 121
Broken Arrow 21, 159
Brown, George 103, 106, 110, 112, 146, 147
Burke Rocker 71

C

Cabarrus County 10, 17
Cameron, James 67, 69, 130
Champion, Douglas 14
Champion, Luveria 147, 149
Champion, Lynwood 147–148
Civil War 11, 17, 18, 37, 57, 58, 60, 62, 136
Cleburne County 10, 19, 25, 30, 60, 67
Cleveland, Heflin 95, 97, 99, 103, 105
Cleveland, Joseph 99, 151
Coker, Johnny 153
Coley, Judge C.J. 10, 13, 74, 138
Confederate gold 57, 58, 59
Confederate States of America 53, 58, 60
copper 19, 37, 58, 60, 61, 65
Cornish miners 60
Cow Pens 67
Creek Cession of 1832 35
Creek Indian removal 35
Creek land 19, 25, 27, 29, 30, 32
Creek Nation 9, 10, 11, 25, 27, 29, 30, 47, 48, 49, 50, 51, 60, 123
Creek Wars 27
cyanide 36, 61, 78, 82, 94, 96, 99

Index

D

Dahlonega 18, 58
Daughtery, Thomas 102, 106, 121
Davis, Jefferson 59
Dean, Lewis 60
Dean, Marvin 153
De Soto, Hernando 15, 16, 32
Devil's Backbone 20, 35, 41, 46, 71
Dr. Ulrich 37, 38
Dutch Bend 38, 67

E

Eagle Creek 35, 76, 146
Ealy Pits 37, 38
Emerson, E.H. 97
Enitachopco 21, 28, 109, 144, 159

F

Fair Labor Standards Act 96
Fanny Goldmine Hill 80
Farrow, Tom 73, 74, 76, 77, 79
Fohsee, Alice 128

G

Gardner, P.S. 92, 93, 94
Georgia 9, 10, 15, 16, 18, 19, 25, 26, 30, 38, 41, 50, 58, 59, 63, 86, 140
Gold City 65
Gold Country 71
Goldville 14, 32, 34, 35, 37, 38, 41, 59, 76, 93, 97, 123, 127, 128, 135, 136, 138, 142
Grayson, George Washington 48, 49
Great Depression 10, 11, 21, 36, 89, 91, 92, 129, 141, 144
Green, Carl 162
Green, Henry Wallace 119
Green, Llewellyn 96, 105, 113
Green, Willie 162
Grierson, Robert 27, 28, 47, 48

H

Hackneyville 36, 58, 69, 123, 125, 129, 136
Haile mine 18
Hawkins, Benjamin 27, 47, 48
Hershel Baker Store 125
Hillabee Blues 58
Hillabee Campground Church 123, 127
Hillabee Creek 21, 28, 34, 36, 37, 38, 123, 125, 128, 144, 159
Hillabee Gold Mining Company 36, 62, 67, 69, 81, 87, 112, 140
Hillabee Town 27, 28
Hillabee tribe 27, 28, 125
Hill, Gregory 46, 71, 72, 76, 77, 78, 79
Hog Mountain 9, 10, 14, 20, 21, 34, 35, 36, 37, 59, 61, 63, 66, 67, 69, 76, 84, 88, 91, 92, 93, 94, 97, 99, 103, 105, 108, 109, 112, 116, 123, 130, 131, 133, 135, 138, 140, 142, 144, 145, 146, 147, 151, 157, 159, 161
Hog Mountain Mining and Milling Company 11, 91, 103, 108, 130
Hooper, Johnson Jones 11, 19, 50, 52, 53, 54, 58, 61, 140
Horseshoe Bend 9, 19, 28
Hot Chapel 128, 129
Houston Mines 136
Hutchinson, Wild Bill 125

I

Indian Removal 9, 18, 20, 29, 51
Indian Removal Act 29, 52
Indian Wars 9, 10, 16, 19, 20, 27, 28

J

Jackson, Andrew 28, 29, 49
Jackson, Kermit 102, 106, 112, 132
Jackson, Talmadge 106, 151, 153
Jarvis, Will 67
Jefferson, Thomas 16, 17, 26

Index

Johnson, James C. 34
Johnson, John 107

K

Kennedy, J.A.P. 85
Kidd, Ben E., III 45, 76, 77
Kidd, Ben E., Jr. 78

L

Lake Martin 35, 41, 71
Leslie, Eugene 131
Log Pit 32, 34, 37, 38
Lovelady, Jesse 147, 162

M

Marable 32
massacres 27
McLeod, Hugh Price 112, 147
McWright, Rhett 106, 155, 158
Mocalumne River 42
Mooney, J.P. 100, 113, 147, 159
Moore, A.H. 37, 63, 65
muckers 100, 102, 153

N

Nelson, Jim 123, 141, 162
Newman, C.E. 129
New Site 100, 109, 123
North Carolina 10, 17, 18, 37, 50, 58, 60, 108, 133

O

Old Federal Road 26, 27
Old Southwest 11, 19, 20, 23, 25, 29, 32, 50, 52, 58, 61, 63
Opothleyahola 50
Osborn, Cecil 153, 155
Osborn, Cowboy 140–142

P

Padgett, Alton 14, 146
Padgett, Elbert 146, 147

panning 38, 62, 79
Park, C.F., Jr. 97
Parmalee, C.H. 71, 72
Patterson, Albert 127, 128, 135, 142
Patterson, Buck 132, 133
Patterson, Delona 127, 137
Patterson, John, governor 10, 13, 127, 135, 137, 142
Patterson, John Love 127, 136
Patterson, Lafayette 97, 128, 136, 138, 142
Peppers, Hershel 113, 149
Philadelphia mint 18
Phillips, James Dowd 20, 41, 42, 45, 46
Phillips, Sarah Ann 20, 41, 42
Phillips, William B. 36, 71, 76
Piedmont region 10, 15, 16, 17, 19, 20, 25, 45, 60
Pine Mountain, Georgia 88

R

Red Stick 27, 28
Reed family 17
Romanoff Land and Mining Company 63, 66
Roosevelt, Franklin D. 92, 96
Russell, Ben 14, 142–143
Russell, Benjamin 94, 142, 143
Russell, Robert 142

S

Saunt, Claudio 49, 50
Savannah & Memphis Railroad 62
Scott, Charlie Thomas 151
secession 57, 58
sharecrop farming 91, 92
Silver Hill Gold Mine 46, 71
Simpson, Daisy 121, 125, 126
Simpson, Hayes 147
Sinnuggee 28, 47
Sky Chief 51
Snow White Sands 59
South Carolina 18, 55, 59, 109, 121, 145
Spaniards 16

INDEX

Street, T.H. 69, 111, 130
Sweet, Palmer C. 17

T

Tallapoosa County 9, 11, 13, 19, 20, 27, 32, 35, 36, 38, 41, 42, 43, 46, 51, 52, 53, 54, 58, 61, 62, 63, 65, 66, 67, 69, 71, 74, 76, 78, 80, 81, 86, 92, 107, 123, 124, 127, 147
Trenholm, George 59
Tuomey, Michael 10, 19, 34, 54, 55, 56, 71

V

Vidette 63, 65
Villa Rica 18, 19, 30
Virginia 15, 16, 17, 59

W

Walls, Marshal Ed 100, 115, 151
Wedowee Schist 15, 25, 97
White, Carey 102, 113, 147, 149
Winn Creek Place 41, 46
Woodruff, Claude 157
World War I 11, 36, 78, 92, 96, 99, 109, 147

Y

Young, Bird H. 11, 52, 53, 54, 61
Young, Otis 23, 46, 146
Youngsville 11, 61

About the Author

Peggy Jackson Walls earned her bachelor of science degree in secondary education at Auburn University in Montgomery and her master of arts in liberal arts, with a minor in southern history, at the Auburn campus. She has taught at Auburn University, Benjamin Russell High School, Central Alabama Community College and the University of Phoenix online. Her article "Gold Mining at Hog Mining in the 1930s" was published by the *Alabama Review* in July 1984. In 1998, she interviewed and wrote the script for a two-hour documentary, *Alexander City: 125 Years of Memories*. She has published articles, poetry and interviews in different mediums.

Her research has been cited in Pulitzer Prize–nominated books *Poor but Proud: Alabama's Poor Whites* (2001) and *Alabama: A History of a Deep South State* (2010). She is the coauthor of *Alexander City* from Arcadia Publishing's Images of America series (2011) and was published in the anthology *Chinaberries and Magnolia Blossoms* from Solomon & George Publishers (2012).

Visit us at
www.historypress.net

This title is also available as an e-book

CPSIA information can be obtained
at www.ICGtesting.com
Printed in the USA
LVHW082139160223
739727LV00003B/87